ADVANCED MATERIALS FOR RENEWABLE HYDROGEN PRODUCTION, STORAGE AND UTILIZATION

Edited by **Jianjun Liu**

Advanced Materials for Renewable Hydrogen Production, Storage and Utilization

http://dx.doi.org/10.5772/59520
Edited by Jianjun Liu

Contributors

Hailiang Chu, Shujun Qiu, Fen Xu, Lixian Sun, Olaitan Lukman Akanji, Ye Mei, Yuriy Zhirko, Jianjun Liu

CBS, Edition **2017**

Published by InTech

Janeza Trdine 9, 51000 Rijeka, Croatia

Publishing Process Manager Iva Lipovic

Technical Editor InTech DTP Team

Cover Designer InTech Design Team

Additional hard copies can be obtained from orders@intechopen.com

Advanced Materials for Renewable Hydrogen Production, Storage and Utilization,
Edited by Jianjun Liu
p. cm.
ISBN 978-953-51-2219-7

INTECH

open science | open minds

Contents

Preface

There is a tremendous lack of public appreciation of scientific and technological difficulties that would be encountered in supplies of energy once fossil fuel sources are acutely depleted. The supply is currently adequate, which allows us to develop alternative energy sources, but the scale of research in physics, chemistry, biology, and engineering needs to be raised, because it will require sustained efforts to solve the long-term problems of global energy security and environmental pollution. Hydrogen is regarded as a suitable energy carrier for sustainable energy schemes with zero emission. In spite of a great deal of scientific and technological efforts, many significant challenges still exist. In 2012, InTech publisher invited me to edit a book titled *Hydrogen Storage* . This book attracted extensive attention from "hydrogen" researchers. This encouraged me to publish a new book with an aim to cover more research fields, namely, hydrogen production, storage, and utilization. As is well known, material issues are a bottleneck in the development of hydrogen economy. Therefore, the *Advanced Materials for Renewable Hydrogen Production, Storage, and Utilization* is defined as our present book title and has tried to present the current research progress in hydrogen applications to the best of our knowledge.

To achieve this, we reviewed all submitted chapters. The chapters cover three research fields: hydrogen production, storage, and utilization. In the past decade, computational materials science, including first-principles calculations, phase field, and finite element analysis, made a significant progress in simulating chemical and physical properties of materials. The Materials Genome Initiative (MGI) involves computation, experiment, and data science and is playing an important role in developing novel materials. This book targets the readership at the professional researcher level. We hope that the book helps in convincing the hydrogen research community of the importance of developing novel hydrogen application materials by combining theoretical and experimental approaches.

The book could have never been realized without the invaluable assistance of Ms. Iva Lipović and Mrs. Ivona Lovrić. Effective communication with them helped to make this book outstanding. I would like to acknowledge the valuable input from my colleague Prof. Jiacheng Wang. I would also like to acknowledge the "One-Hundred-Talent Project," the "Key Research Program (Grant No. KGZD-EW-T06)" of the Chinese Academy of Sciences, and the research grant (no. 14DZ2261200) from the Shanghai government, China, for their financial support. Most importantly, without the love and encouragement of my family, I could not have embarked on or finished this book project.

Jianjun Liu

State Key Laboratory of High Performance Ceramics and Superfine Microstructure

Shanghai Institute of Ceramics

Chinese Academy of Sciences

China

Section 1

Introduction

Chapter 1

Introductory Chapter

Jianjun Liu

Additional information is available at the end of the chapter

http://dx.doi.org/10.5772/61943

1. Introduction

1.1. Importance of developing hydrogen energy

As the global consumption of fossil fuels is growing at an alarming and unsustainable rate, the associated emissions of greenhouse gases and other toxic pollutions are reaching levels that are environmentally unacceptable. A large challenge of world economic development requires meeting the demand of energy consumption while reducing emissions of greenhouse gases and pollutants. Since the sun is our largest and cheapest energy resource, the utilization of solar energy is considered as the ultimate strategy for solving this problem. It continuously bombards our planet with solar energy, with 1 h of solar energy equating to more than all of our annual energy consumption [1]. Among the limited methods for solar energy conversion and utilization, solar water splitting to make hydrogen has been considered as one of the most effective and cleanest ways. Hydrogen, as an energy carrier, is widely regarded as a potential cost-effective, renewable, and clean energy alternative to petroleum, especially in the various transportation tools [2]. The hydrogen produced by solar energy can then be used by fuel cells to generate electricity with zero-emission of water. Figure 1 displays the scheme of renewable energy based on some selected technologies including hydrogen production, storage, and utilization.

Although a great deal of efforts have been made to develop a sustainable hydrogen economy, many challenges still exist in many research fields such as hydrogen generation, storage, and fuel cells in the cyclic system. In order for solar energy to be the major contributor to generate clean hydrogen, the conversion efficiencies of solar water-splitting devices need to be improved, for example, developing highly active catalysts for hydrogen/oxygen evolution reactions. One significant barrier that prevents the widespread use of hydrogen as an energy source is the lack of efficient hydrogen storage systems. For on-board application, the hydrogen storage system is required to provide the needed quantity of hydrogen with accepted

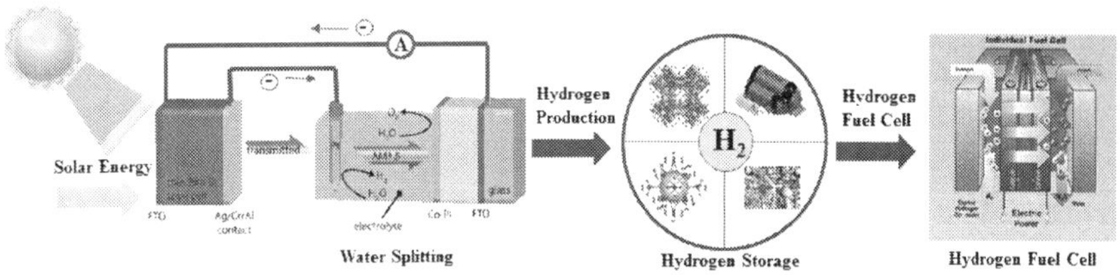

Figure 1. A scheme of renewable energy (solar energy to hydrogen) based on some selected technologies including hydrogen production, storage, and utilization.

engine. The currently available hydrogen storage systems include compressed hydrogen, nanomaterials, metal hydrides, and complex metal hydrides. Each can meet some of the requirements but none of them meets all the requirements for critical applications. Therefore, developing a compact and efficient hydrogen storage technology is the most technically challenging for achieving hydrogen economy. Hydrogen fuel cells are a pathway to convert hydrogen-carried energy into electricity. The development of high-performance fuel cells is leading to higher efficiencies and cleaner energy utilization than the other conventional technologies. Costs vary over a wide range, and future cost projections depend on mass production estimates, driven by the automobile market, in volumes and quantities that do not represent the volumes of trucks, buses, and trains.

Advances of materials science have generated a significant impact on many technology fields [3]. Scientists always think, work, and interact to develop and optimize advanced materials and process in different time and space scales. Developing high-performance materials is playing an important role in creating new fields of fundamental sciences and promoting practical technologies. In order to achieve a technological breakthrough in effective renewable energy conversion from solar energy to hydrogen energy, we need to develop advanced materials which can be applied in some technologies of hydrogen production, storage, and utilization. To this end, the potential capabilities of advanced high-performance materials must be extensively discovered at a more fundamental level.

The specific aim of this book is to provide a fundamental study and in-depth understanding of mechanisms involved in hydrogen applications. In the past years, there are a few review books to discuss development of advanced materials in hydrogen production, or hydrogen storage, or hydrogen fuel cells. However, a comprehensive review in these three aspects is not found so far. Herein, we intend to provide a complete review in the present and previous research progresses including discovered materials and research methods in experiment and computation.

2. Advanced materials in hydrogen production, storage, and utilization

Nano-photocatalysts for hydrogen production plays an important role in enhancing conversion efficiency from solar energy to hydrogen chemical energy [4]. The highly active catalysts should have suitable band gaps which are favorable to improve the absorbance of abundant solar light to drive hydrogen and oxygen evolution reactions. In addition, good charge-transfer ability for electrons and holes and high surface catalytic reactivity for half-reaction are very important for this kind of photocatalysts.

Loading cocatalysts onto photocatalysts to form hydrogen or oxygen evolution sites have been considered as an effective method to enhance photocatalytic activity for water splitting. In the past decades, different kinds of materials such as transition metals and transition metal compounds (oxides, carbides, nitrides, and sulfides) have been developed as effective cocatalysts for photocatalytic water splitting [4, 5] Some noble metals such as Pt, Ru, Au, Rh, etc., and metal oxides such as NiO_x, Cr_2O_3, etc., perform well for water reduction cocatalysts by entrapping electrons from semiconductor. In addition, some noble metal oxides such as IrO_2 and RuO_2 as well as spinel and layer materials such as Co_3O_4 and $LiCoO_2$ have been found to have effective water oxidation by entrapping holes [6].

Solid-state nanomaterials for hydrogen storage perhaps offer the best opportunity for meeting the requirements of on-board applications. Nanoscience and nanotechnology involve studying and working with matter on a nanometer scale. Nanomaterials with at least one dimension <100 nm can be in the form of thin films or surface coatings, nanowires and nanotubes, or nanoparticles. The unique properties of nanomaterials originate from the increased surface area and quantum effects, both of which relate to the small size. Therefore, reactivity, strength, and electrical characteristics of nanomaterials may be significantly different from those of conventional materials. For example, a particle 30 nm in size has only 5% of its atoms on its surface. When the size is reduced to 3 nm, the particle has 50% of its atoms exposed on the surface. This makes nanoparticles have a much greater surface area per unit mass than bulk structures. Both surface area and exposed atoms on the surface could be useful for storing hydrogen. Therefore, nanotechnology is expected to play a key role in designing high-capacity solid-state hydrogen storage materials. In fact, the concept of nanoscience and nanoengineering has been actively exploited in improving existing hydrogen storage materials and searching for new hydrogen storage candidates. For example, nanocatalysts have been used to improve kinetics of hydrogen uptake and release or to improve hydrogen storage capacity through spillover [7]. Novel building blocks have been proposed to maximize the hydrogen capacity and optimize the strength of hydrogen binding. Large-surface-area nanomaterials which offer more host atoms/sites for hydrogen and allow easy access of these sites have been synthesized, as in materials such as molecular organic frameworks (MOFs) and covalent organic frameworks (COFs), and show promise as hydrogen storage media [8, 9].

The photoelectrochemical (PEC) water splitting has been considered as the most attractive method over other hydrogen production approaches. The pioneering work of solar-driven water splitting in PEC frame consisted of TiO_2 anode and Pt cathodes for oxygen and hydrogen evolution reaction [10]. In fact, TiO_2 represents one of the most important semiconductor materials for PEC water splitting. However, due to its wide band gap, TiO_2 cannot absorb visible and infrared light for solar water splitting. Doping of many metals and non-metals has been widely adopted to narrow the band gap [11]. In the past decades, many different nanoengineering methods such as tailoring morphology and composition and interface have been applied to improve the performance of light absorption of TiO_2.

ZnO, another widely used semiconductor, has also been widely investigated as a photoanode of PEC water splitting. In the past years, different approaches including ion doping and visible light sensitization with narrow band gap semiconductors have been used to expand the light absorption region and hence improve the PEC water-splitting performance [12, 13]. Hematite (α-Fe_2O_3) is an increasing promising material for solar-driven PEC water-splitting applications due to its relatively narrow band gap (2.1 eV). However, the poor charge-transfer property of α-Fe_2O_3 greatly prevents its efficiency as a photoanode material for splitting water. Similarly, ion doping and nanocomposition have been extensively applied to improve its charge transfer property [14].

Nanostructured electrocatalysts for hydrogen fuel cells can harness the chemical energy of hydrogen to convert electricity at a high-energy conversion efficiency of >70% without combustion and pollution, owing to the slow kinetics of oxygen reduction reaction on the cathode, which is primarily responsible for reduced voltage and low conversion efficiency. The noble metal Pt has been regarded as the most active electrocatalyst for this reaction. However, the high cost and scarcity of Pt greatly limit the development and widespread commercial application. In the past two decades, many different catalysts such CrN, WC, and Mo_2C have been studied to replace Pt [15]. Although none of them has a comparative catalytic activity as Pt, their resource is abundant and available at low cost.

3. Outline of the book

The primary focus of the book is on developing advanced materials for hydrogen production, storage, and utilization. Unravelling structure-related chemical/physical properties of these materials plays an important role in designing high-performance materials. The chapter is concluded with a descriptor of chemical and physical properties of materials.

Chapter 2 presents electron-microscopic, energy-dispersion, and optical (IR-absorption and low-temperature (4.5 K) photoluminescence) investigations of InSe- and GaSe-layered crystals intercalated using the below considered methods with various concentrations of hydrogen or hydrogen-containing molecules such as water, toluene, and alcohol. These materials have promising applications of these layered matrices (and powders from them) when creating operation elements of solid hydrogen accumulators. At the same time, these InSe- and GaSe-layered crystals themselves with different types of conductivity and their thin-layer intercalates as well as other similar compounds can be considered as promising candidates for creation of various hybrid-layered structures for flat flexible displays. Chapter 3 presents the extensive studies for understanding hydrogen bonding interaction mechanism in biomolecules, which is favorable to develop hydrogen production technology by biomass approach.

As an example for hydrogen storage, hydrogen desorption/adsorption processes in a metal hydride bed reactor are simulated using the software COMSOL Multiphysics to study the diffusion and heating of hydrogen and metal hydride powder in both radial and axial directions. They are presented in Chapter 4. The model consists of a system of partial differential equations (PDE) describing two-dimensional heat and mass transfer of hydrogen in a porous matrix. The influence of the operating parameters such as temperature, pressure, concentration, permeability, and thermal conductivity on the rate of absorption/desorption of hydrogen in metal hydride will be fully discussed. The simulation results obtained could be applied to the on-board hydrogen storage technology, in particular for the hydrogen supply of a fuel cell for powering of a hydrogen fuel cell vehicle.

Chapter 5 reports a research progress on hydrogen storage based on metal amidoboranes and their derivatives. Extended from ammonia borane as high-capacity hydrogen storage materials, thermodynamic tailing of dehydrogenation of metal amidoboranes, metal borohydride–ammonia borane complexes, and metal amidoboranes ammoniate as well as their derivatives was studied with emphasis on syntheses, crystal structures, and dehydrogenation properties. These studies clearly are greatly important to further design novel high-capacity hydrogen storage materials. In addition, the authors also summarize nanoconfinement and nanocatalysis of ammonia borane and metal amidoboranes, and their derivatives.

Author details

Jianjun Liu*

Address all correspondence to: jliu@mail.sic.ac.cn

State Key Laboratory of High Performance Ceramics and Superfine Microstructure, Shanghai Institute of Ceramics, Chinese Academy of Sciences, Shanghai, China

References

[1] L. Vayssieries, *On Solar Hydrogen and Technology*, John Wiley and Sons, Singapore, 2009.

[2] S. S. Mao, S. Chen, *International Journal of Energy Research* 31 (2007) 619–636.

[3] F. D. S. Marquis, *Journal of the Minerals Metals and Materials Society* 63 (2011) 43.

[4] X. Chen, S. Shen, L. Guo, S. S. Mao, *Chemical Reviews* 110 (2010) 6503–6570.

[5] S. Shen, J. Shi, P. Guo, L. Uo, *International Journal of Nanotechnology* 8 (2011) 523–591.

[6] S. Shen, S. S. Mao, *Nanophotonics* 1 (2012) 31–50.

[7] Y. W. Li, R. T. Yang, *Journal of the American Chemical Society* 128 (2006) 726–727.

[8] N. L. Rosi, J. Eckert, M. Eddaoudi, D. T. Vodak, J. Kim, M. O'Keeffe, O. M. Yaghi, *Science* 3 (2003) 1127–1129.

[9] S. S. Han, H. Furukawa, O. M. Yaghi, W. A. Goddard, *Journal of the American Chemical Society* 130 (2008) 11580–11583.

[10] A. Fujishima, K. Honda, *Nature* 238 (1972) 37–38.

[11] S. U. M. Khan, M. Al-Shahry, W. B. Ingler Jr., *Science* 297 (2002) 2243–2245.

[12] A. I. Hochbaum, P. Yang, *Chemical Reviews* 110 (2010) 527–546.

[13] Y. J. Lin, G. B. Yuan, R. Liu, S. Zhou, S. W. Sheeha, D. W. Wang, *Chemical Physics Letters* 507 (2011) 209–216.

[14] K. Sivula, F. Le Formal, M. Gratzel, *ChemSusChem* 4 (2011) 432–449.

[15] H. Zhong, X. Chen, H. Zhang, M. Wang, S. S. Mao, *Appllied Physics* Letters 91 (2007) 163103.

Section 2

Hydrogen Production

Chapter 2

Characterization and Optical Properties of Layered InSe and GaSe Crystals Intercalated with Hydrogen and Hydrogen-Containing Molecules

Yuriy Zhirko, Vasiliy Grekhov, Nikolay Skubenko,
Zakhar Kovalyuk and Taras Feshak

Additional information is available at the end of the chapter

http://dx.doi.org/10.5772/61051

Abstract

Described in this review are our electron-microscopic, energy dispersion, and optical (IR absorption and low-temperature (4.5 K) photoluminescence) investigations of layered InSe and GaSe crystals intercalated using the considered below methods with various concentrations of hydrogen or hydrogen-containing molecules (HCM) of water, toluene, and alcohol. It has been shown that these crystals are efficiently intercalated–deintercalated (IC–DIC) not only with hydrogen but also with HCM. It has been ascertained that the layered crystals kept for a long term in natural conditions contain molecules of water and carbon dioxide gas in the subsurface area, where their concentration increases with time. It has been shown that the concentration of molecules intercalated into the crystal decreases when deepening inside its bulk both along crystal layers and the normal to them. Also adduced were empirical equations relating concentration distribution of molecules intercalated into the crystal with the duration of IC–DIC processes. Ascertained are the optimal dimensions allowing to efficiently perform IC–DIC of the whole crystal bulk. It has been shown that the IC–DIC process in time has the look of a hysteresis loop—fast initial growth of the HCM concentration with the following slowdown when intercalating and fast drop with the following retardation when deintercalating under permanent evacuation of the chamber with the crystal. The availability of two types of conductivity in n-InSe and p-GaSe crystals causes some features in the emission spectra of the crystals intercalated with hydrogen or HCM. In the case of n-InSe crystals, the availability of HCM results, in the whole, in predominant emission of free excitons, and in the opposite case of p-GaSe crystals, the recombination of carriers via deep centers becomes predominant.

Keywords: Layered crystals, intercalation, SEM, EDX, photoluminescence

1. Introduction

Unique chemical bonds—strong ion-covalent hybrid$s$-p^3-bonds inside crystalline layers and weak van der Waals ones between layers, which take place in layered InSe and GaSe single crystals, provide the possibility to introduce foreign atoms and molecules into their interlayer space by using intercalation. From the practical viewpoint, these crystals and their intercalates are promising materials for solid-state hydrogen accumulators, electric power, and solar cells as well as for numerous applications in other tasks.

As it was shown in [1], the electrochemical intercalation of hydrogen into InSe and GaSe crystals results in changing their physical properties. In particular, spectra of exciton absorption in these crystals demonstrate two-dimensional localization of exciton movement inside the plane of crystal layers, and NMR spectra show resonance absorption of electromagnetic radiation by hydrogen molecules available both in the interlayer and intralayer spaces of the crystals. In this case, the chemical shift of the energy level inherent to molecular hydrogen in these crystals depends both on the hydrogen concentration and crystal temperature, while the intensity of resonance absorption bands for this hydrogen depends on direction and value chosen for applied magnetic field relatively to the plane of crystal layers. Calculations performed using the model of 2D and 3D localization of hydrogen molecules in the layered crystals showed [1] that the processes of intercalation and deintercalation of molecular hydrogen take place with participation of full-symmetric optical vibrations in the crystal lattice, which modulate the width of potential wells and barriers. Results of investigations [1] also showed that the hydrogen concentration in these crystals reaches considerable values comparable with the concentration of molecules in the very matrix and in powders of the crystals the amount of hydrogen molecules after the sufficient duration of intercalation can exceed the amount of matrix molecules by 2–2.5 times.

The aim of this work was to study changes of physicochemical properties arising in InSe and GaSe crystals in the course of intercalation with hydrogen, water, and more complex hydrogen-containing molecules of alcohol and toluene, as well as to obtain information about spatial distribution of the concentrations inherent to intercalated molecules in these crystals and dynamics of their deintercalation from the crystals by using the methods of optical and electron microscopy, energy dispersion, VIS, and IR spectrometry. In Section 2, we have adduced some (necessary to discuss the results obtained in this work) information about InSe and GaSe crystals, their physicochemical properties, and ways for the intercalation of simple nonorganic and organic molecules into these layered crystals. Section 3 describes results of our optical-microscopic, electron-microscopic, and energy dispersion investigations of InSe and GaSe crystals intercalated with hydrogen-containing molecules of water, alcohol, and toluene.

Adduced in Section 4 are the results of investigations concerning GaSe crystals kept for more than 30 years in natural conditions of ambient medium ("aged" crystals). Concentrations and chemical composition of molecules embedded from ambient medium in the process of natural intercalation into the subsurface area of the aged crystals are estimated there. Also, the estimates of spatial distribution inherent to simple nonorganic molecules in the bulk of the layered crystals within the volume scales of > 1 cm^3 down to < $1 \bullet 10^{-10}$ cm^3 are made in both 2D and 3D versions. Besides, the results of investigations aimed at infrared light absorption by the aged GaSe crystals within the range of valence vibrations intrinsic for molecules of water and carbon dioxide are discussed; adduced are the analytical expressions relating the dependences of water molecules concentration in these crystals with the duration of their intercalation–deintercalation; given are recommendations concerning the spatial dimensions of the crystals from the viewpoint of their efficient application in hydrogen energetics.

Section 5 of the work describes the results of low-temperature investigations aimed at photoluminescence spectra of InSe and GaSe crystals intercalated with hydrogen molecules as well as molecules of water, alcohol, and toluene (containing OH and CH groups) with the purpose to ascertain the influence of their intercalation (including their passivation) on optical properties of these crystals within the range of the fundamental electron (exciton) transition. In summary, using the examples of InSe and GaSe crystals, we have presented generalized conclusions about the character of processes related with the intercalation–deintercalation of simple nonorganic and organic molecules into the layered crystals as well as their influence on physicochemical properties of these crystals in the context of applying these layered matrixes (not containing carbon) in hydrogen energetics.

2. Experimental methods

InSe and GaSe layered crystals considered in this chapter belong to binary compositions A3B6, in which crystalline layers are composed of four monolayers of Se and In(Ga) atoms, located one above another, in the sequence Se-In(Ga)-In(Ga)-Se and united together by ion-covalent hybrid sp^3-bonds. Inside these crystalline layers, the orientation of atoms correspond to the spatial group D_{3h}. In this case, each In(Ga) atom inside the crystalline layer provides the three nearest Se atoms with two own valence electrons to fill their *p*-shells, while one remaining electron together with another excess electron of the adjacent In(Ga) atom forms the completed *s*-shell. As a result, orbitals of valence electrons are localized inside the layer and practically do not overlap with orbitals of valence electrons from neighboring layers. As a consequence, there is a weak molecular van der Waals bond between these crystalline layers. This clearly pronounced anisotropy of chemical bonds between crystalline layers as well as inside these layers allows the intercalation of layered crystals, i.e., introduction of foreign atoms or molecules into interlayer space of a grown crystal. For instance, in the case of InSe and GaSe crystals, the volume of van der Waals space (the so-called "van der Waals gap") is equal to 40–45% of the total crystal volume, and the internal surface of this space is close to (2–2.5) 10^3 m^2 in 1 cm^3.

It is well known [1–4] that InSe and GaSe single crystals grown by the Bridgman method have four polytypes (β, δ, ε, and γ) that differ between each other by the sequence of packing the crystalline layers. As a rule, InSe single crystals have the γ-polytype with rhombohedral (trigonal) crystalline structure (spatial group C^5_{3v}, the primitive unit cell of which contains one

In_2Se_2 that comprises three layers). The nonprimitive hexagonal unit cell comprises three crystalline layers and consists of three molecules In_2Se_2. GaSe single crystals have ε-polytype. They belong to the hexagonal crystalline structure (spatial group D^1_{3h}), the unit cell of which consists of two Ga_2Se_2 molecules located in the framework of two crystalline layers.

Parameters of the crystalline lattice inherent to γ-InSe and ε-GaSe crystals are well investigated. In the case of ε-GaSe, they are as follows: the lattice parameter along the direction perpendicular to the layers is C_0 = 15.95 Å, and inside the layer plane it is a_0 = 3.755 Å [5]. For γ-InSe, they are, respectively, C_0 = 25.32 Å and a_0 = 4.001 Å. The distances between the nearest atoms inside the layer are as follows: $C_{In\text{-}In}$ = 2.79 Å and $C_{In\text{-}Se}$ = 2.65 Å, angle <φ = 119.3°, layer thickness C_l = 5.36 Å, distance between layers C_i = 3.08 Å, and the distance between the nearest Se atoms in adjacent layers $C_{Se\text{-}Se}$ = 3.80 Å [6].

Undoubtedly, the unique properties of this class of semiconductor compounds, in particular, InSe and GaSe crystals, attract special attention of researchers because the heterostructures based on them not only possess high photosensitivity and can be applied in solar cells [7–9] and accumulators of electric energy [10] but are also rather promising for the creation of γ-radiation sensors, as it follows from [11]. At the same time, as it was shown in recent studies [1, 12–14] that the layered InSe and GaSe crystals can be applied in hydrogen energetics as operating elements in solid hydrogen accumulators. The hydrogen concentration in powders of these crystals can reach values close to x = 5–6, where x is the amount of embedded hydrogen atoms per one formula unit in the matrix of intercalated crystal.

Figure 1 shows volume images of molecules related to hydrogen, water, alcohol, toluene, or carbon oxide gas that were embedded into InSe and GaSe crystals by intercalation. The following methods were used in these cases: electrochemical intercalation, intercalation under pressure when working with hydrogen, or natural immersion deepening the samples to solutions of water, alcohol, and toluene, or long-term (for 30 years) keeping the samples in ambient atmosphere.

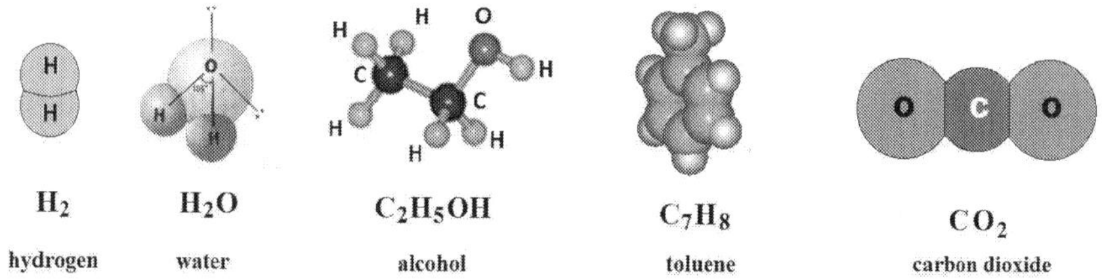

Figure 1. Volume images of molecules, namely, hydrogen, water, alcohol, toluene, and carbon oxide gas.

As seen from Figure 1, in their composition, majority of molecules have hydrogen. At the same time, the molecules of water and alcohol contain the hydroxyl group O-H, the molecules of alcohol and toluene–carboxyl group C-H, and the molecule of carbon oxide –C=O group.

http://dx.doi.org/10.5772/61051

2.1. The method of electrochemical intercalation

The method of electrochemical intercalation with using "pulling" electric field is presented in Figure 2. It was used when intercalating the crystals with hydrogen. Previous studies [12, 13] investigated layered InSe and GaSe crystals chemically intercalated with hydrogen, where the authors used the one-chamber three-electrode cell [1, 15] (see Figure 2a). When current passes through the cell, the composition of solution is changed due to redox processes taking place both on operation and on auxiliary electrodes. Products of reactions that go at the auxiliary electrode get to the zone of operation electrode and can interact with it, which results in changing its potential and can lead to incorrect determination of the embedded intercalant concentration.

Therefore, in this work, we used the cell separated by porous membrane spaces of operation and auxiliary electrodes (Figure 2b). The application of this cell enabled us to separate processes of oxidation and reduction and to prevent mixing solutions contacting with operation and auxiliary electrodes.

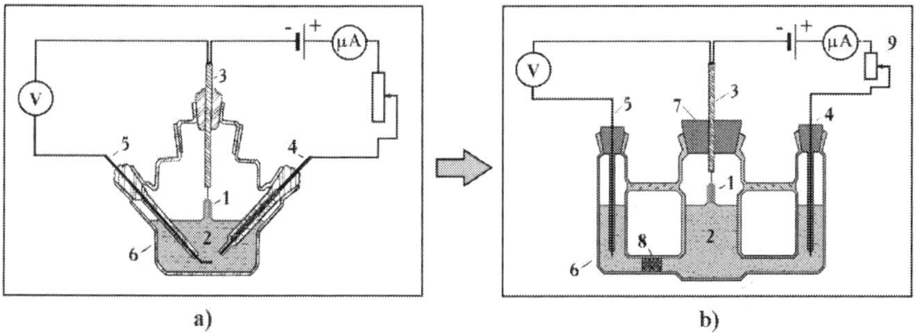

Figure 2. (a) One-chamber cell; (b) separated cell. The scheme of the electrochemical cell for the intercalation of layered InSe and GaSe crystals with hydrogen. 1—sample of InSe or GaSe (operation electrode, OE) 2—electrolyte; 3—operation electrode; 4—chlorine-silver reference electrode, CE; 5—auxiliary electrode (AE); 6—case of the cell; 7—hole for the sensor of pH-meter; 8—porous membrane; 9—potentiostate ПИ-50-1.

It is noteworthy that the electrochemical cell used in the process of intercalating the layered crystals with hydrogen should satisfy the following requirements:

- Material of the cell should not interact with electrolyte. In our case, the cell was made of silica glass.
- Cell should be hermetic. It is acceptable to use Teflon, vinyl plastic, or chemical-resistant rubber.
- Space between operation and auxiliary electrodes should be separated with a porous membrane, for example, with the glass Schott filter.
- Cell design should provide measurements of the electrolyte pH level during the intercalation process.

When carrying out electrochemical intercalation, InSe and GaSe single crystals were used as the operation electrode. Their samples had the shape of parallelepiped. In accord with the method [16], thin copper wire was soldered to the sample from side opposite to that contacting with electrolyte. Another end of this wire was taken off through the cork and added to the potentiostat. As a reference electrode, we used the chlorine-silver electrode of industrial production. A thin platinum wire soldered in glass served as an auxiliary electrode.

To prepare electrolyte, we used concentrated hydrochloric acid of the "chemically pure" grade diluted in bidistilled water. H ions were embedded into the layered InSe and GaSe crystals as a result of cathode polarization in galvanostatic conditions (i_a = 50–100 μA/cm^2) from 1 N HCl water solution under potentials of the operation electrode that did not allow release of gas-like hydrogen. The process was carried out like that used in electrolysis bath under permanent control of the operation electrode potential with using the chlorine-silver electrode deepened into 1 N HCl solution. In the course of intercalation, the electrolyte pH level was measured using the pH-meter HI-8314 through the special hole in the cell cork. The fact of intercalation with hydrogen ions was confirmed using chemical analysis.

The concentration of embedded hydrogen was determined via the amount of electrical charge passed through the cell, i.e., using the controlled parameters of intercalation process: the current density (j) and the duration of the process (t). This process is described by the following relation [17]:

$$C_0(i,t) = A \cdot e^{[\alpha(t)+\beta(t)]}, \tag{1}$$

where $C_0(i, t)$ is the amount of embedded intercalating ions per unit volume of the crystal; $\alpha(t)$, β(t) are some linear functions of embedding process duration; i is the current passing through the cell; and A is the coefficient of proportionality.

The influence of embedded protons concentration on properties of indium monoselenide was studied using the same group of samples by their step-by-step additional intercalation. This process was accompanied by a shift of the value of equilibrium electrode potential inherent to initial material. Taking into consideration the fact that the stationary potential E for the initial nonintercalated indium selenide is close to 0.25 V, and data of chemical analysis show that the higher hydrogen content is corresponded by the higher negative E value, one can draw a conclusion that the obtained shift of the equilibrium potential after intercalation is related with the creation of hydrogen compounds embedded.

2.2. Natural intercalation of InSe and GaSe single crystals

In this work, we also investigated intercalation processes in InSe and GaSe single crystals under pressure 2 to 5 atm in sealed silica ampoules filled with molecular hydrogen, or using the natural way by deepening the crystals into chemically pure solutions of water, alcohol, and toluene. In addition, we studied the samples of GaSe crystals that were kept for 30 years in ambient atmosphere. To perform physicochemical investigations of naturally intercalated single crystals, from monocrystalline InSe and GaSe ingots, we cut discs and then prepared plates with the thickness 3 to 5 mm and lateral dimensions 5 × 5 mm^2.

Natural intercalation with hydrogen under pressure was realized within the range of 2 to 5 atm in hermetic cells at the temperature 200°C for 3–4 hours. The natural immersion intercalation of the samples with molecules of water, alcohol, or toluene was performed at room temperature in silica ampoules by deepening them into chemically pure solutions of corresponding substances for 35 and 165 days. After finishing the process, the intercalated samples of InSe and GaSe were withdrawn from solutions.

2.3. The method for the deintercalation of hydrogen from layered crystals

Hydrogen deintercalation from electrochemically intercalated layered H_XInSe and H_XGaSe crystals were performed for 3–9 hours at T = 110°C in vacuum under constant pumping out. It was ascertained that the degree of deintercalation in the H_XInSe samples practically linearly grows with the hydrogen concentration from 60% for $x \to 0$ up to 75–80% for $x \to 2$. When increasing the hydrogen concentration up to $x \to 4$, the degree of deintercalation in H_XInSe and H_XGaSe crystals grows up to 85%. Investigation of powder-like InSe and GaSe crystals showed that the hydrogen concentration in them can reach the values x = 5–6, while the degree of deintercalation can reach 90%. Deintercalation of gases from the crystals kept for 30 years in natural conditions was carried out inside a vacuumed chamber. This method will be described below, when discussing the results of EDX investigations.

3. Physical properties of InSe and GaSe crystals intercalated with hydrogen and hydrogen-containing molecules

As it was shown by the previous electron-microscopic investigations of InSe and GaSe crystals electrochemically intercalated with hydrogen [1], even the 80,000-fold magnification, in the whole range of investigated hydrogen concentrations x from 0 up to 4, where x is the amount of molecules per one molecular unit of the matrix, electron images of crystalline surfaces have no formations of structural defects related with embedding hydrogen into these crystals. At the same time, small-angle XRD investigations showed [1, 14] that, with increasing the concentration of intercalated hydrogen, as a result of intercalated hydrogen pressure on the matrix, there grows deviation of matrix parameters in intercalate H_XGaSe from classical values in GaSe crystals (hexagonal group 9,6/mmm). In particular, for H_XGaSe crystals at T = 300 K and x = 1.0, the parameter C_0 is increased by 0.031 ± 0.003 Å from C_0 = 15.94 Å up to C_0 = 15.971 Å, and the parameter a_0 by 0.006 ± 0.003 Å from a_0 = 3.753 Å up to a_0 = 3.759 Å.

It is noteworthy that, despite essential deviation of the lattice parameter after hydrogen intercalation from classical values, electron-microscopic images of crystal surfaces remain practically invariable even for high hydrogen concentrations, and any additional formations of nonuniformities on surfaces or inside the samples are not observed. At the same time, as it was ascertained in [1], the samples of GaSe crystals contain residue of the monoclinic phase of red β-Se, and dimensions of these inclusions do not exceed 1 mkm. The inclusions disappear after annealing of the crystals in a silica ampoule for 3 hours in vacuum at the temperature T = 350°C.

Investigation of EDX spectra taken from surfaces of InSe and GaSe crystals performed using the scanning microscope Zeiss EVO 50 XVP equipped with the detector INCA ENERGY 450 showed that chemical compositions of these crystals are as follows: 49.3% Se atoms and 50.7% In atoms in InSe crystals as well as 50.08% Se atoms and 49.92% Ga atoms in GaSe crystals.

3.1. Crystals intercalated with hydrogen-containing molecules

To study the influence of intercalation with hydrogen-containing molecules (HCM) of water, alcohol, and toluene on the crystals InSe and GaSe, we carried out optical, electron-microscopic (SEM), and energy-dispersed (EDX) investigations of the samples intercalated with HCM for 35 and 165 days. To reach it, we prepared six samples of InSe and the same amount of GaSe crystals intercalated naturally with molecules of above substances for 35 and 165 days, respectively, as well as reference samples of nonintercalated crystals and those electrochemically intercalated with hydrogen. When performing this part of work, we used electron microscopes of the firm TESCAN-SEM VEGA 3, SEM MIRA 3 with an immersion magnetic lens, two-beam SEM LIRA 3 equipped with energy-dispersed EDX spectrometers X-Max or BRUKER, as well as optical transmission microscope Primo Star 5 of the firm Carl Zeiss.

As can be seen in Figure 3, after the intercalation of the crystals with HCM (in particular, GaSe crystals with water for 165 days) even for 1,000-fold magnification in the optical microscope, one can observe the formation of a bubbled surface. More clearly pronounced such a surface can be seen in Figure 4 when using 50,000- or 100,000-fold magnification in the electron microscope SEM MIRA 3 with an immersion magnetic lens.

It is interesting that the same changes with the formation of bubbles after intercalation with water take place in InSe crystals, too. However, when intercalating the samples InSe and GaSe with alcohol or toluene, the formation of bubbled surfaces is essentially less pronounced as compared to the case of water.

It should be noted that under low magnification (Figures 5–7), sample surfaces with clear images of bubbles border with other surfaces where one can observe quite different formations (see light areas). For the case of 30,400-fold magnification, these areas are represented in Figure 6 as a color version. As can be seen, under this magnification, they take a look of some bushes or dendrites located at the crystal surface.

Demonstrated in Figure 7 as an example is the look of the surface for sample 2 (GaSe, H_2O, 35 days) obtained in the regime BSE + SE using SEM VEGA 3 with magnification 5,000. As one can see, after intercalation with water for 35 days, the sample surface is not so developed with dendrites as compared to that after long-term intercalation.

To ascertain the nature of these dendrites, we perform additional SEM and EDX investigations. Shown in Figure 8 are SE and BSE images of the GaSe crystal surface after intercalation with

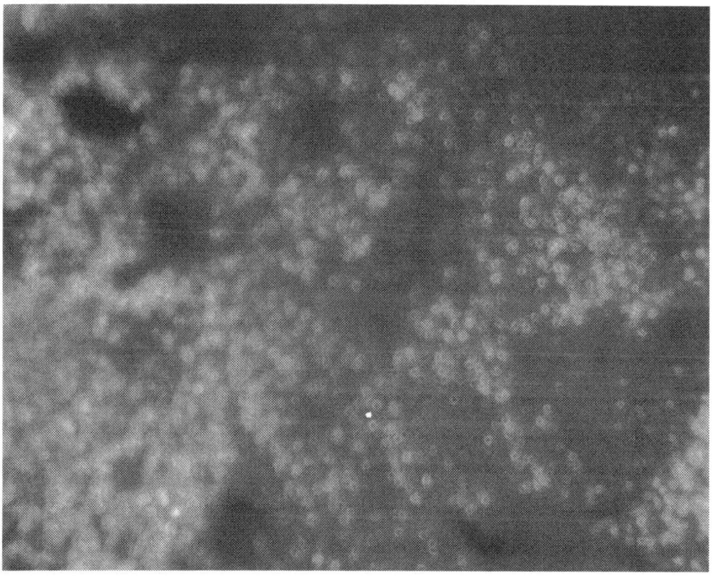

Figure 3. Optical transmission images of the GaSe crystal before (a) and after (b) its intercalation with water for 165 days, obtained using the transmission microscope Primo Star 5 at 1,000-fold magnification.

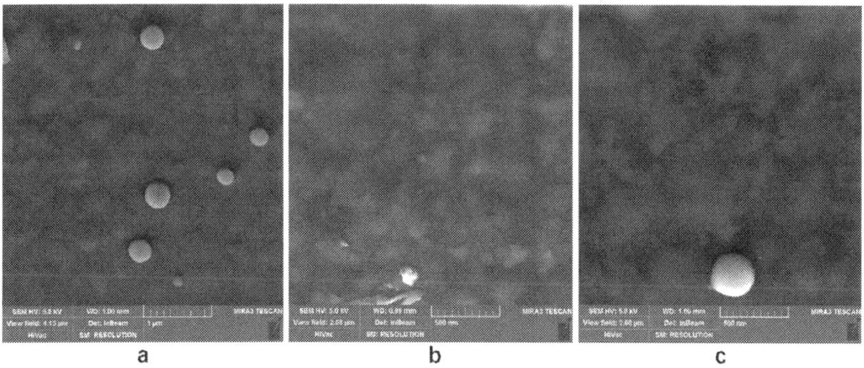

Figure 4. Obtained using SEM MIRA 3 images of several points on the sample surface for the crystal GaSe (H_2O, 165 days) with bubbles and remains of the β-Se phase for the accelerating voltage 5 kV: (a) point 1, magnification 50,000, scale—1 mkm; (b) point 2, magnification 100,000, scale—0.5 mkm; (c) point 3, magnification 100,000, scale—0.5 mkm.

water for 165 days. Seen there are the dendrites after processing them by using the two-beam electron microscope SEM LYRA 3. Lateral dimensions of the hollow etched out by the second beam are close to 9 × 14 mkm, and its depth is 1.5 mkm. It is clearly seen that these dendrites grows through the surface from the crystal bulk, while their "roots" remain inside the crystal.

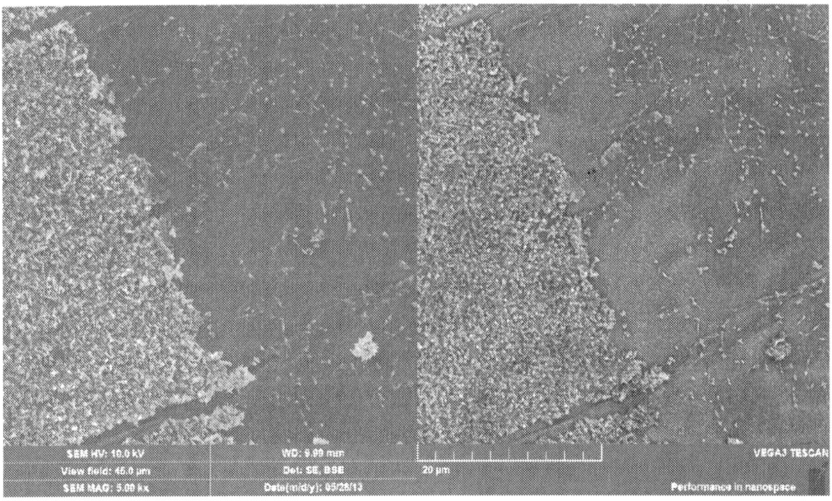

Figure 5. BSE + SE images for the surface of the sample GaSe (H_2O, 165 days) obtained using SEM VEGA 3 with magnification 5,000 and accelerating voltage 10 kV. Scale—20 mkm.

Figure 6. BSE + SE images for the surface of the sample GaSe (H_2O, 165 days) obtained using SEM VEGA 3 with magnification 30,400 and accelerating voltage 30 kV. Scale—5 mkm. Color—after computer processing.

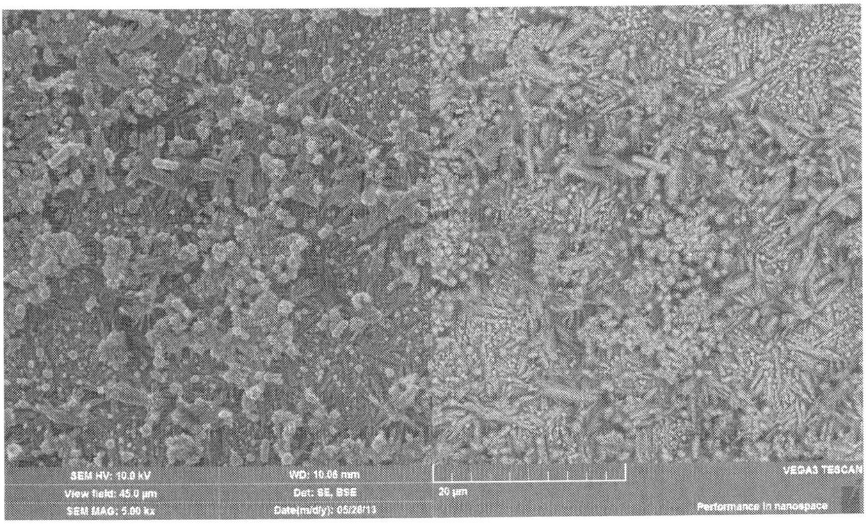

Figure 7. BSE + SE images for the surface of sample 2 (GaSe, H_2O, 35 days) obtained using SEM VEGA 3 with magnification 5,000 and accelerating voltage 10 kV. Scale—20 mkm.

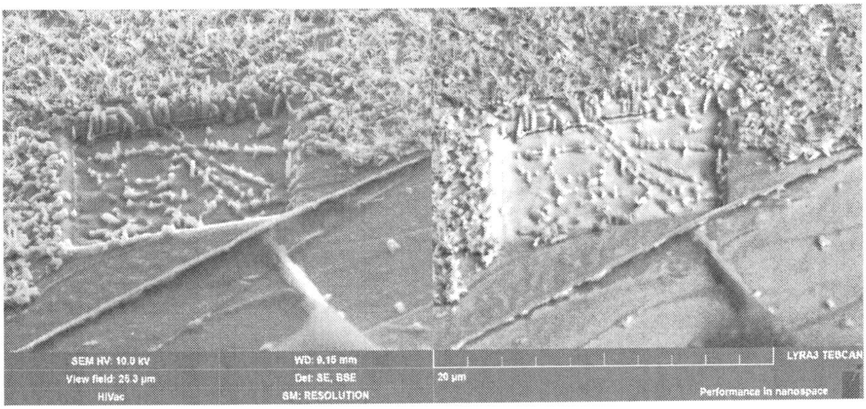

Figure 8. SEM LYRA 3, magnification 5,000. Scale—20 mkm.

To further ascertain the chemical nature of these dendrites, we use the same microscope SEM LYRA 3. It enables us to choose one of the dendrite samples that remained on the surface after cutting the hollow. Using the Ga-ion gun, this sample was fixed on the probe (see Figure 9). Then the end of the probe was attached to the sample holder, soldered to it by using the same gun and cut from the probe (Figure 10).

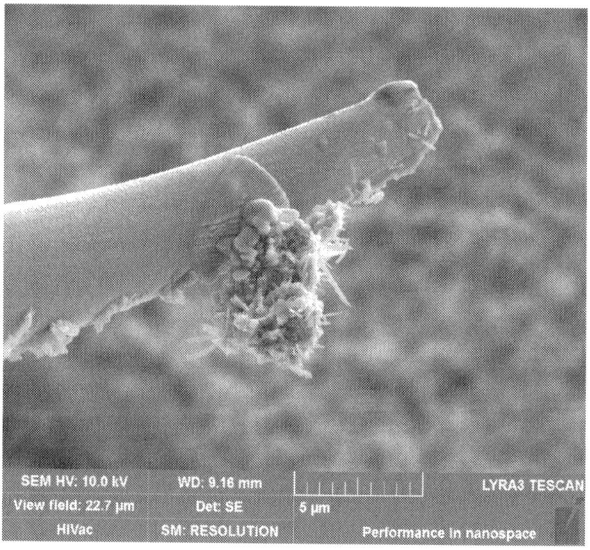

Figure 9. Spatial image of the dendrite fixed on the probe in SEM LYRA 3. Scale—1 mkm. Accelerating voltage 10 kV.

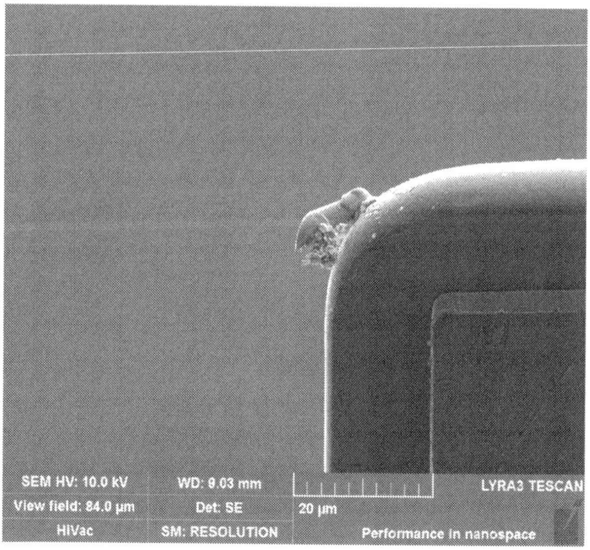

Figure 10. Spatial image of the residue of probe with the dendrite part fixed on the holder in SEM LYRA 3. Scale—20 mkm. Accelerating voltage 10 kV.

http://dx.doi.org/10.5772/61051

The dendrite sample on the probe, obtained in such a manner for EDX investigations, was transferred to SEM MIRA 3 equipped with the immersion magnetic lens. As seen in Figure 11a, for 150,000-fold magnification the dendrite structure is rather complex in 3D space. Under 500,000-fold magnification and the slope of the subject table 40° relatively to Z-axis, shown in Figure 11b is the image of one pin of the dendrite. It is clearly seen that the shape of the pin is close to cylindrical or cone-like. Its diameter at the end reaches 25–50 nm and is narrowed with the step 15 ± 2 nm. The same electron microscope SEM MIRA 3 enabled us to perform EDX investigations and ascertain that these dendrites mainly consist of Se atoms and, as seen from Figure 8, grow from the foreign phase of remaining β-Se.

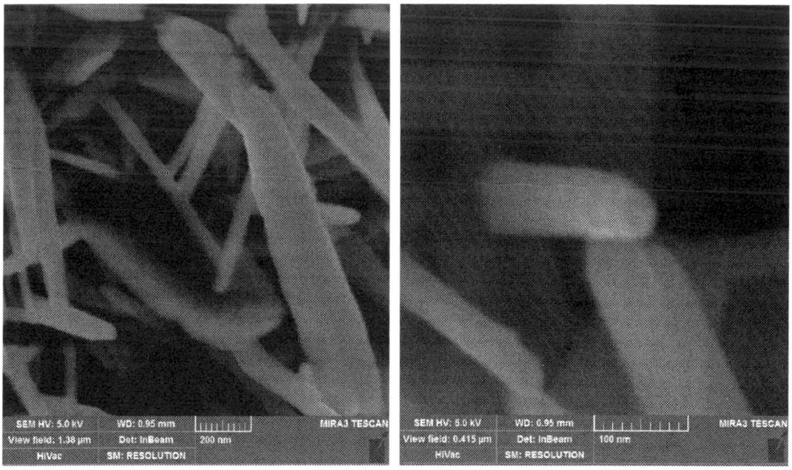

a b

Figure 11. SE images of surfaces in sample 1 (GaSe, H_2O, 165 days) taken from areas with dendrites for very small WD. Microscope SEM MIRA 3, accelerating voltage 5 kV; magnifications: 150,000-fold (a) and 500,000-fold (b), scales—200 and 100 nm, respectively.

Energy dispersion investigations of GaSe crystal surfaces after water intercalation for 165 days also enabled to ascertain (see Tables 1 and 2) that in areas of dendrite localization, the oxygen concentration is approximately 3 times higher than in areas of minor amount of dendrites. When water molecules are present on the surface of crystals (in particular, GaSe), it can indicate the formation of developed complexes with β-Se accompanied by Ga_2O_3 crystallites. As it follows from the tables, the increase in the Se concentration in these areas also correlates with these conclusions.

It should be noted that the formation of dendrites at the presence of water is more clearly pronounced in GaSe crystals than in the InSe ones, which is caused by different types of conductivity in these crystals. In accord with [18], it is dominance of acceptors over donors in GaSe crystals that causes precipitation of residual β-Se phases as well as binding and layering the crystalline layers during growth of these crystals. The results of performed investigations allow concluding that the formation of bubbled surfaces in the course of intercalation with HCM is caused by partial layering along the van der Waals gap and filling this space of the crystal with molecules of water, alcohol, or toluene. It is seen from the example of GaSe that lateral dimensions of bubbled formations are commensurate with inclusions of residual β-Se in nonintercalated crystals and lie within the range 0.1 to 1.0 mkm.

Element	**Series**	**Unn. [wt. %]**	**C norm. [wt. %]**	**C Atom [at. %]**	**C Error (1 Sigma) [wt. %]**
Dendrites					
Carbon	K-series	0.00	0.00	0.00	0.00
Oxygen	K-series	6.16	9.52	33.15	1.03
Gallium	K-series	20.63	31.86	25.46	0.63
Selenium	K-series	37.96	58.62	41.38	1.23
	Total :	64.75	100.00	100.00	
Bubbles					
Carbon	K-series	0.00	0.00	0.00	0.00
Oxygen	K-series	2.22	2.80	11.85	0.47
Gallium	K-series	33.94	42.77	41.51	0.98
Selenium	K-series	43.19	54.43	46.64	1.38
	Total :	79.35	100.00	100.00	

Table 1. Distribution of chemical elements over the areas with dendrites and bubbles in sample 1

IC duration		**HCM concentration**	
H_2O	C_2H_5OH	C_7H_6	
35 days	12%	6.0%	5.2%
165 days	25%	12,5%	9,9%

Table 2. HCM concentration in GaSe and InSe intercalates

Similar energy dispersion investigations were carried out using the crystals InSe and GaSe intercalated with toluene and alcohol for 35 and 165 days. The obtained concentrations of oxygen atoms after intercalation with water, of oxygen and carbon atoms after intercalation with alcohol, and of carbon after using toluene enabled us to estimate concentrations of HCM molecules as well as their dependences on the time of intercalation. The results of calculations shown in Table 2 confirm that the HCM concentration in the layered InSe and GaSe crystals grows by two times, when the duration of intercalation is increased from 35 to 165 days. It allows making the first estimate of the dependence of the HCM concentration in these crystals on the time of deepening the sample in the respective solutions of HCM by using the following simple analytical expression

$$A(t) = A_0 \cdot \sqrt{A_0 \cdot t}, \tag{2}$$

where $A(t)$ is the concentration of molecules intercalated into crystal, A_0 is the fitting parameter for the crystal and solution, and t is the time of intercalation.

3.2. Conclusions

The performed optical, electron-microscopic, and energy-dispersed investigations enabled authors to ascertain that nonintercalated crystals do not possess stoichiometric composition. For instance, the chemical composition of crystals InSe and GaSe has the following atomic ratio: 49.3% Se atoms and 50.7% In atoms in InSe crystals as well as 50.08% Se atoms and 49.92% Ga atoms in GaSe. It is well correlated with their conductivity, namely, n-type for InSe and p-type for GaSe. The amount of residual β-Se in the interlayer space of GaSe crystals is much higher than that in the InSe ones.

It has been ascertained that the concentration of water, alcohol, and toluene molecules even in the case of the natural intercalation of InSe and GaSe crystals with time can reach considerable values ($x \to 1$) commensurate with those after electrochemical intercalation.

It has been found that

i. molecules containing hydroxyl group (OH) are embedded into InSe and GaSe crystals better than molecules containing the only carboxyl group (CH);

ii. contrary to the case of hydrogen, high concentrations of HCM (when $x \to 1$) lead to the blistering of the crystals and formation of bubbles in the interlayer space, which are filled with separate HCM or their liquids;

iii. long-term intercalation with HCM (especially with water) results in the formation of dendrites on the surface of InSe and GaSe crystals, which have a look of bushes consisting of red monoclinic β-Se that forms (accumulates) in the process of synthesis of these crystals. Geometrical dimensions of these bushes can reach 2 to 5 mkm. Their pins have round or oval shape and diameter within the range 25–50 nm, they are narrowed to their ends with the step 15 ± 2 nm;

iv. since the amount of the residual β-Se in InSe crystals is lower than that in GaSe, the amount of formed dendrites in them is lower, too;

v. in the areas where dendrites accumulate, one can observe an increased concentration of HCM.

3.3. Acknowledgement

The authors express their sincere gratitude to the firm TESCAN for the given possibility to carry out investigations by using the up-to-date samples of electron microscopes as well as to its co-workers Ph.D. Martin Petrenec, Michal Svoboda, Jiri Dlugos for their assistance in performing the part of this work, when studying the crystals InSe and GaSe intercalated with hydrogen-containing molecules.

4. Electron microscopy and energy dispersion investigations of GaSe crystals kept in natural ambient conditions

Presented in this chapter are the results of SEM and EDX investigations aimed at percentage and distribution of atmosphere gases in GaSe crystals kept for more than 30 years in natural ambient conditions of the medium geographic belt—in summer at temperatures up to +35°C, in winter down to –35°C, and air humidity from 30% up to 100%. With this aim, we chose to cut 30 years ago flat monocrystalline GaSe plates of the 12-mm width and 6-mm thickness, which were used for preparation of the samples with dimensions 10 × 12 mm and thickness 6 mm (in what follows, the crystals of the first type). Then the results of studying them were compared with those inherent to the GaSe crystals of the second type—as-grown crystals nonintercalated and intercalated with water. In detail, the crystals of the second type were described in the previous chapter.

4.1. SEM and EDX investigations

Energy dispersion (EDX) investigations of the atmosphere gases content in the crystals of the first type at surfaces containing planes of crystalline layers were performed using the scanning electron microscope SEM VEGA 3 (TESCAN). As these crystals are semiconductors, the investigations were performed without deposition of conducting layers on substrates, in the regime of medium vacuum, when the residual pressure in the column and chamber was close to $2{\bullet}10^{-2}$ Pa. Registration of secondary electrons was made using the standard SE detector. Spectra of γ-radiation were measured using the energy-dispersed analyzer X-Max (OXFORD).

The optical image of the first type of GaSe crystal with the thickness 100 to 150 mkm obtained using the transmission optical microscope Primo Star 5 with 1,000-fold magnification is shown in Figure 12. It is seen that, in GaSe crystals of the first type as well as in those of the second type intercalated with water (see Figure 3 in Chapter 2), volume images contain nonhomogeneities that cannot be observed in as-grown crystals.

EDX investigations enabled to ascertain that the stoichiometric composition of Ga and Se atoms in the first type of crystals is the same as in as-grown crystals of the second type (studied in detail in [1]). At the same time, the results of these surface investigations showed the following:

1. There is a great amount of atomic oxygen in the crystals of the first type (see data in Figure 14d) as well as in the crystals of the second type intercalated with water.

http://dx.doi.org/10.5772/61051

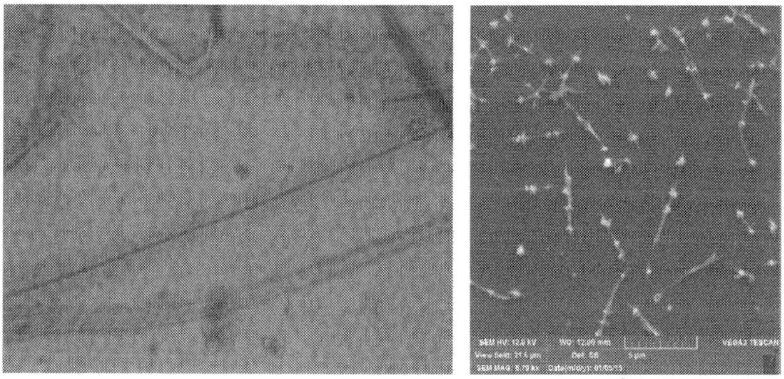

a b

Figure 12. (a) Optical image of the first type of GaSe crystal obtained using the transmission optical microscope Primo Star 5 with 1,000-fold magnification. (b) SE image of the surface obtained for the same crystal under accelerating voltage 12 kV and magnification 8,790 times. Scale—5 mkm.

2. On surfaces of crystals of the first type (Figures 12b and 12a) as well as in the crystals of the second type intercalated with water (see Figures 4 and 5 in Chapter 2), one can observe nuclei of residual β-Se phase and dendrites of monocrystalline β-Se.

3. In the bulk of the first type of crystals (Figure. 13b, fresh cleavages at the depth close to 1 mm from the surface), contrary to the second type of as-grown crystals, these nucleation phases of residual β-Se cannot be already observed.

a b

Figure 13. SE image for small WD taken: (a) from the crystal surface and (b) from the cleavage at the depth 1.1 mm, obtained at the accelerating voltage 5 kV and magnifications ×44,100 and ×40,300, respectively. Scale—1 mkm.

Figure 14 shows the SE image of the first type of GaSe crystal under accelerating voltage 12 kV and magnification ×2,000. Marked with the point in Figure 14a is the area used for EDX analysis. The spectrum of X-ray radiation corresponding to transitions of electrons between inner shells of atoms and analysis of chemical elements made in the virtual program shell INCA are represented in Figure 14c and in Figure 14d, respectively. It should be noted that the results adduced in Figures 13a and 13b were obtained in the first day of investigations.

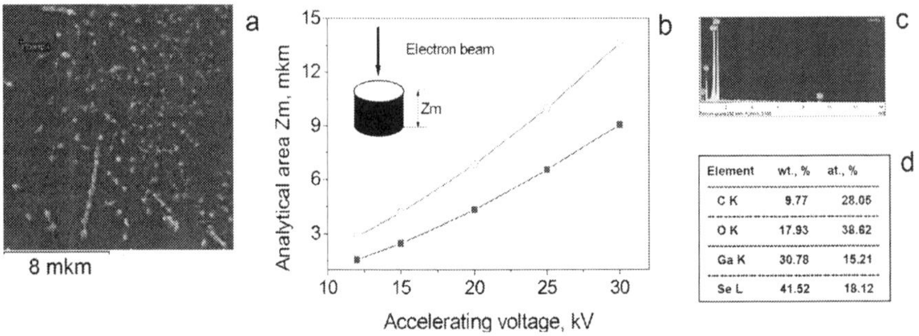

Figure 14. (a) Part of the first type of GaSe crystal and the point in it where EDX analysis was performed; (b) dependence of the size of analytical area Z_m (see the insert) on the accelerating voltage of the probing electron beam, which was obtained using Eq. (3.3) for oxygen (circles) and carbon (squares); (c) EDX spectrum of the point selected in Figure 14a; (d) chemical composition of elements in point 1 shown on a Figure 14a.

As it is seen from Figure 14d, in the points (and areas) containing no dendrites and residual phases of β-Se, surfaces of the first type of crystals contain a large amount of carbon and oxygen atoms. It is known that the Earth atmosphere mainly consists of molecules of nitrogen (78.1%), oxygen (21), argon (0.9%), water vapors (0.5–1.5%), and carbon dioxide (near 1%). Since, beside carbon and oxygen, other chemical elements present in the gas composition of the Earth's atmosphere were not found on the surface of the first type of GaSe crystals, it allows assuming that natural subsurface areas of these crystals as well as corresponding areas of the second type of crystals intercalated with water can contain a large amount of water molecules and carbon dioxide from ambient atmosphere.

Unfortunately, the data available in this chapter do not allow separating the intrinsic amount of carbon atoms in crystals of the first type from carbon available in the column and chamber of the microscope. At the same time, we can assume that the first type of crystal surfaces contain water and carbon dioxide molecules in approximately the same amounts. It permits us to exclude carbon from further discussion and to focus on the availability of oxygen. In this case, it is not excluded the possibility that subsurface area of the crystals in the presence of water molecules and molecular oxygen is subjected to substitution of Se atoms with the oxygen ones with the creation of a dielectric coating Ga_2O_3 in accord with the scheme

$$2GaSe + 3H_2O \Rightarrow Ga_2O_3 + 3H_2 + 2\beta\text{-Se}, \quad (3)$$

or

$$2GaSe + 3O_2 \Rightarrow Ga_2O_3 + 2\beta\text{-}Se. \quad (4)$$

Previous investigations of the second type of crystals are indicative of the fact that the process given by Eq. (3) is more preferable on the crystal surface in presence of water since it allows explaining the growth of dendrites from monocrystalline β-Se in the areas of nucleation phases of residual β-Se and probably decreasing amount of residual β-Se in the bulk of first type of crystals. The process given by Eq. (4) in our opinion was happened in a subsurface crystal area. Despite the described chemical processes of Ga_2O_3 formation in the crystals of the first and second types, we assume that the main contribution to EDX spectra corresponding to presence of oxygen is conditioned by molecules of water and carbon dioxide embedded in the process of natural (atmospheric or immersion) intercalation into the layered crystal. In this case, as it follows from data on a Figure 14c, up to two atoms of oxygen are available per one molecule of Ga_2Se_2 in the matrix of the first type of crystals.

Since GaSe crystals of the first type contain a large amount of oxygen atoms from the composition of H_2O and CO_2 molecules, the authors performed temporal investigations of the processes related with the deintercalation of these molecules from the crystals. Every time after performing the EDX investigations, the first type of crystals remained in the pumped out chamber of microscope for 1 week. For this time, the residual pressure in the chamber increased from 2×10^{-2} up to 2×10^{2} Pa till the next day after measurements and up to 3.5×10^{2} Pa till the end of the week. Further, the next cycle of measurements was carried out.

Thus, seven cycles of measurements were repeated with the 1-week interval between them. The results of EDX investigations (by using the accelerating voltages for electrons from 15 to 30 kV) are adduced in Figure 15a, while their averaged voltage values are summarized in Table 3. Also, for comparison, presented in this table are the results of previous EDX investigations aimed at the second type of crystals after their intercalation with water for 35and 165 days.

K_0 - averaged concentrations of water molecules per one formular unit of GaSe crustal in subsurface regionв

1-st type					2-nd type	
1 day	1 week	3 weeks	5 weeks	7 weeks	35 days	165 days
108 %	78 %	67 %	33 %	13 %	12 %	25 %

Table 3. The percentage of water determined using the EDX spectra of water intercalates of the layered $GaSe[H_2O]_X$ crystals of the first and second types, where X is the amount of water molecules per one formula unit of the crystal

The performed EDX investigations of the first type of crystals under the values of accelerating voltages for electrons from 15 to 30 kV showed that, with increasing Z_m—depth of penetration typical for probing electron beam (calculated for oxygen atoms)—from 3.5 up to 13 mkm, the parameter K_0—concentration of oxygen atoms—decreases in the crystal. The increase in the duration of evacuation also results in decreasing the oxygen concentration in the subsurface layer (see Table 3 and Figure 15b).

These results are shown in Figure 15a with open squares, rhombs, and circles for the cases of 7-, 21-, and 35-day keeping the first type of crystal in vacuum. The skew cross designates the water concentration in the subsurface area in the first day of investigations under the accelerating voltage 12 kV. It should be noted that the analytic area Z_m of the probing beam for chemical elements (in this case: gallium, selenium, oxygen, and carbon) was determined using the adopted fitting formula

$$Z_m = 0.033 \cdot (E_0^{1.7} - E_C^{1.7}) \cdot \frac{A}{\rho \cdot Z}, \tag{5}$$

where E_0 is the accelerating voltage (kV), E_C is the minimum band emission energy (keV), A is the atomic mass, ρ is the density (kg/m^3), and Z is the atomic number.

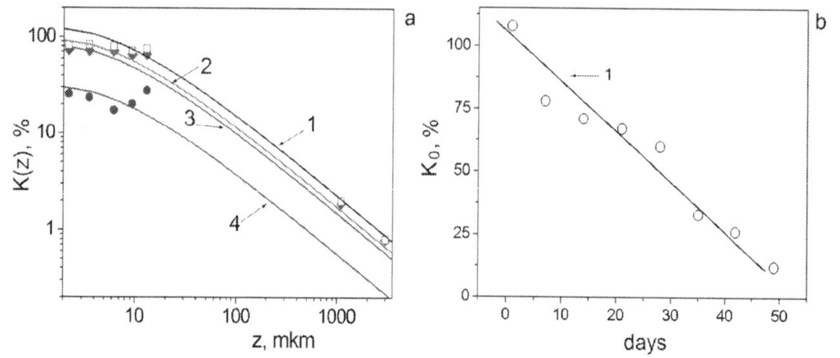

Figure 15. (a) Squares, rhombs, and circles correspond to experimental values of the water concentration in the subsurface area of the first type of crystal versus the depth of penetration typical for the probing electron beam in the case of 1-, 3-, and 5-week crystal being in vacuum, respectively. Crosses designate experimental values of the water concentration in the first day of measurements at the depth 2.24 mkm as well as on the fresh cleavage at the distance 1,100 mkm from the crystal surface and in the middle of the crystal at the depth 3,000 mkm obtained from IR investigations. The curves 1–4 correspond to fitting 2D dependences of the water concentration versus the crystal depth. The dependences were obtained in accord with Eq. (6) for the case of semi-infinite crystal. (b) The dependence of the fitting parameter K_0 on the duration of keeping the first type of crystal in vacuum for the case of semi-infinite 2D crystal. Curve 1 is approximation of the parameter K_0 dependence on time.

As seen from Figure 15a, with increasing the volume of the analytic area Z_m, the concentration of oxygen atoms in each of these measurements drops with increasing the depth of probing electron beam penetration into crystal. It allows ascertaining the relation between the parameter K_0—concentration of water on the surface of layered crystal and parameter K(z)—its distribution along the depth in the crystal bulk (z) in the direction perpendicular to crystal layers in the following form

$$K(z) = \frac{e \cdot K_0 \cdot \ln(z+1)}{z+1}, \tag{6}$$

where e = 2.718... is the Napierian base. Note that this dependence is obtained for the case of semi-infinite crystal.

As seen from Figure 15a, the curves 1 to 3 both qualitatively and rather well quantitatively describe experimental values of the water vapor concentration K(z) in the crystal of the first type as a function of z-distance into the crystal bulk from its surface at fixed K_0 values obtained after 1-, 3-, and 5-week being in vacuum. In Figure 15a, curve 1 plotted for the first day of measurements is indicative of a fast drop in the concentration of water molecules with depth. It is also seen (Figure 15a, curve 4) that, with increasing the duration of evacuation, the experimental values of the water vapor concentration K(z) in the crystal bulk begin to decline from the theoretical ones. It confirms that water and carbon dioxide molecules faster go out from subsurface areas and slower from the crystal bulk.

The similar phenomena of decreasing the water concentration are observed on the fresh cleavages of the first type of crystals (see Figure 13b) when shifting the point of measurements from an edge to the center. For instance, for the cleavage at the distance z = 1100 mkm from the surface, the concentration of oxygen (measured immediately after obtaining the fresh cleavage) is decreased from 5.5% to 4% on the area 500 × 500 mkm located in the center of the sample down to 1.5–1.1% on the area 50 × 50 mkm in the same center. It indicates that in the first type of crystals, the concentration of intercalated water and carbon dioxide gas is also decreased in the lateral plane from the layer edge to its center. These results are shown in Figure 13a, with the asterisk for the case of the first day measurements on the area 50 × 50 mkm at the value z = 1100 mkm.

Thus, one can assume that the concentration of H_2O and CO_2 molecules in the layered crystal decreases in the crystal bulk when moving from the surface into the bulk to its center—along the perpendicular and in the plane of crystal layers. Shown for clearness in Figure 16a is the 3D sketch of concentration distribution for H_2O and CO_2 molecules in the layered crystal with increasing the depth—along the perpendicular (z-axis) and in the plane of layers (x, y axes).

We suppose that this simultaneous decrease in the concentration when shifting into the bulk as well as to the center of the sample is caused by the decrease in the amount of H_2O and CO_2 molecules penetrating along the layers and perpendicularly to them at presence of covalent intralayer and van der Waals interlayer forces. Like to the case of molecular hydrogen intercalation in GaSe and InSe [1], this intercalation of H_2O and CO_2 molecules takes place due to participation of A'_1-full-symmetric intralayer vibrations of Ga_2Se_2 molecules.

These A'_1 vibrations at room temperature are responsible for changing the width of potential barriers (interlayer Ga–Se bound) and the widths of 2D (van der Waals gaps) and 3D (interlayer space) quantum wells where the localization of molecules takes place. In the case of H_2O molecules, the localization of water vapors results in the formation of bubbles inside the

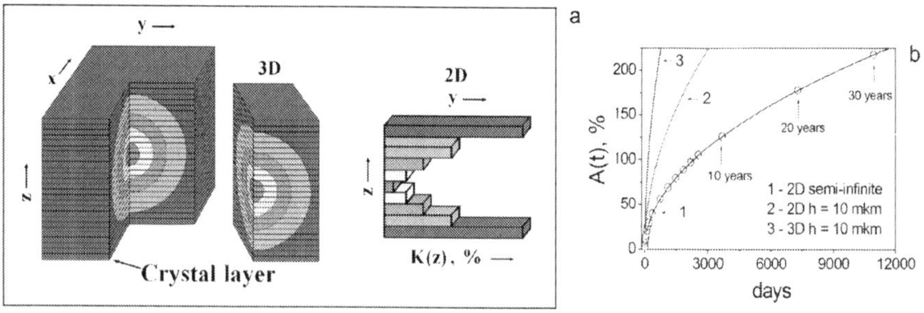

Figure 16. (a) The 3D sketch of distribution inherent to the concentration of water molecules in the layered crystal within the plane (x, y) and along the perpendicular (z) to crystal layers. Also shown here is the 2D crystal cut in the plane (y, z). (b) Dependence of the water molecule concentration on the surface of the second type of crystal on the duration of the intercalation process. The circles and curve 1 correspond to the case of semi-infinite 2D distribution in accord with Eq. (2); curve 2—for the "thin" crystal with the thickness h = 10 mkm; curve 3—for 3D crystal with the cube face h = 10 mkm.

interlayer space, as it was shown in the previous chapter. The similar bubbles are also seen in subsurface areas of the first type GaSe crystals.

Thereof, being based on Eq. (5) describing the distribution of water molecules along the direction perpendicular to crystal layers (when the crystal is semi-infinite) and on the sketch for the generalized 3D distribution of the concentration inherent to H_2O and CO_2 molecules inside the crystal GaSe, one can estimate the distribution of the concentration K(z) for these molecules in the case of a real crystal with the thickness $z = h$. Here, Eq. (5) should be transformed to the look

$$K(z)_{2D} = \frac{e \cdot K_0 \cdot \ln(z+1)}{(z+1)} + \frac{e \cdot K_0 \cdot \ln(h-z+1)}{(h-z+1)}, \quad (7)$$

which reflects the 2D cut of concentration distribution in a three-dimensional crystal. The integrated value of the molecular concentration for the 2D cut shown in Figure 16a at the fixed crystal thickness $z = h$ can be defined as

$$K_{2D} = \int_0^h K(z)d(z). \quad (8)$$

Since in the first approximation one can assume that the distribution K(z) does not depend on direction within the plane (x, y), the integrated value of specific molecule concentration in the bulk crystal of a cubic shape with faces h should have the following look

$$K_{3D} = \left(K_{2D}\right)^{3/2}. \tag{9}$$

In accord with Eq. (8), Figure 17 shows the function for 2D distribution of concentrations for H_2O and CO_2 molecules and various values of the crystal thickness h. For convenience, summarized in Table 4 are also the numeric values of the fitting parameter K_0 on the surface as well as integrated values for the concentration of water molecules K_{2D} in the 2D cut and K_{3D} in 3D crystal bulk as a function of dimensions inherent to crystal faces (x, y, z).

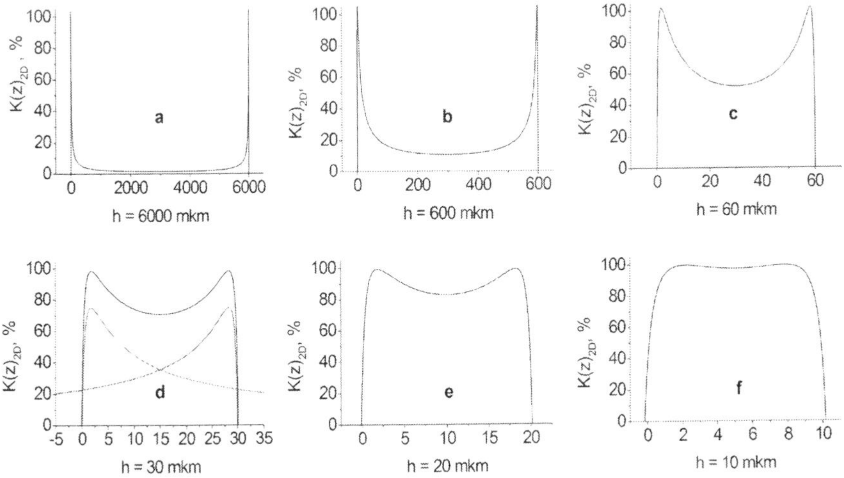

Figure 17. The 2D cut distribution of the concentration $K_{2D}(z)$ in the case of infinite crystal with the thickness $z = h$.

As seen from Figures 17a to 17f, with decreasing the thickness of crystal, the integrated concentration of molecules K_{2D} in the 2D cut (area under the curve) increases from 3.4% for h = 6000 mkm up to 94% for h = 10 mkm. In this case, the integrated concentration in the bulk K_{3D} (see Table 4) also increases from 0.63% up to 91%. In addition, the contribution from opposite faces becomes more essential due to this decrease of geometrical dimensions of the crystal. In accord with Table 4, it results in a practically 2-fold decrease in the value of K_0 parameter that defines the concentration of water molecules on the front face from $K_0 = 100\%$ for $h = 600$ mkm down to $K_0 = 52\%$ for $h = 10$ mkm. For clearness, this case is adduced in Figure 17d as superposition of distribution functions K_{2Z} from the opposite crystal faces.

In summary of this discussion, we shall perform juxtaposition of results obtained after EDX investigations concerning deintercalation (in vacuum) of the first type samples with the similar data obtained in the process of immersion intercalation of thesecond type samples for the case of water molecules.

Length – (mkm): 2D case x = y = ∞, z = h; 3D case x = y = z = h	Adjusting concentration of K_0 on a surfaces for 2D case, (%)	Concentration K_{2D} in case of 2D cross-section, %	Concentration K_{3D} in 3D case, %
6000	100	3.4	0.63
600	99	18.5	38
60	90	67	55
30	80	80	72
20	70	88	83
15	66	92	88
10	60	94	91
5	52	91	87

Table 4. Values of the fitting parameter K_0 and specific concentration K_{2D} on thickness ($z = h$) for the 2D cut ($x = y = \infty$), and for the specific concentration K_{3D} for a cube with face dimensions $x = y = z = h$

4.2. The model of the process for the intercalation–deintercalation of water molecules from GaSe crystals

In accord with the conclusions obtained in Section 3, the concentration of water intercalated into the crystal GaSe of the second type is described with the Eq. (2).

Let us consider that the parameters A_0 and K_0 describing processes of intercalation and deintercalation of water in GaSe crystals of the first and second types in the crystal subsurface area are compatible with each other. They unambiguously describe the process of intercalation–deintercalation of water in layered crystals. (Note that in the case of the second type of crystals, geometrical dimensions of crystals intercalated with water were 5 × 5 × 1 mm^3.) It follows thereof that in both cases of EDX investigations, the semi-infinite method of intercalation–deintercalation of surfaces took place. Therefore, contribution to the fitting parameters A_0 and K_0 from the opposite faces can be neglected. Then the processes of intercalation–deintercalation in the crystal semi-infinite by its thickness are identical to each other and can be described as follows

$$A(t,z) \equiv K(z,t). \tag{10}$$

where the functions $A(t)$ and K(z) have now the look $A(t,z) \equiv K(z,t) = f(z,t)$. Then Eqs. (6) and (2) can be transformed to the look

$$K(z,t) = \frac{e \cdot K_0(t) \cdot \ln(z+1)}{z+1}, \quad a$$

and (11)

$$A(t,z) = A_0(z) \cdot \sqrt{A_0(z) \cdot t}. \quad b$$

Thereof, if using the fitting parameter value $A_0 = 1.63 \pm 0.5$ obtained to describe water intercalation into semi-infinite crystal of the second type, it follows that $A(t)$ in the subsurface area will reach the value equal to 100% in $t = 6.5 \pm 0.5$ years. After keeping the sample in water for 30 years, in accord with Eq. (2) and curve 1 in Figure 14b, $A(t)$ will reach the value 215 ± 5%, which is 2-fold higher than after keeping the sample in ambient atmosphere.

Combining Eqs. (2), (8) and Eqs. (2), (9) with Eqs. (11a), (11b), one can prove that with decreasing the thickness of quasi-infinite crystal down to $h = 10$ mkm (when the parameter value of $K_0 = 50\%$ is sufficient), the function $A(t, z)$ (curve 2 in Figure 14d) will reach 100% in 600 ± 20 days, and when the crystal will be shortened to the cube with faces of 10 mkm, 100% intercalation will occur in 150 ± 5 days (curve 3, Figure 14b).

Figure 18 shows a generalized process of intercalation–deintercalation of water in crystals GaSe for the 3D case, when $x = y = z = h = 10$ mkm. Curve 1 is plotted in accord with Eq. (11b), and curve 2 is in accord with Eq. (11a). Experimental points correspond to the values shown in Figure 13b.

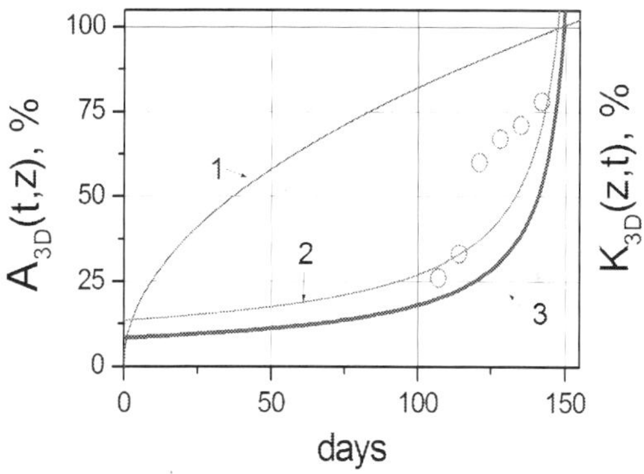

Figure 18. Scheme for the process of water intercalation–deintercalation into GaSe crystal. Curve 1 illustrates the process of intercalation in the cubic crystal with the face 10 μm, and curves 2 and 3 describe the process of deintercalation from the infinite cubic crystal of the thickness 10 μm and cubic crystal with the face 10 μm, respectively. Open circles show experimental values $K_0(t)$ on the surface of the crystal semi-infinite by its width.

As seen from Figure 18, the curve of intercalation–deintercalation looks like a hysteresis loop —a rather long process of natural intercalation with water is followed by rather fast deintercalation in vacuum under permanent pumping out. It is seen that the initial fast process of deintercalation at high concentrations is changed for the slow one at low concentrations. To provide full evacuation of water vapors, it needs period at least three times longer. It should be noted that curves 1 to 3 in Figure 18 only qualitatively describe these processes. In practice, these intercalation–deintercalation processes in crystals with small dimensions of faces are considered by us theoretically and need additional experimental investigations.

Since similar dependences of the water concentration on the duration of intercalation were also obtained for InSe crystals, we believe that the character of water intercalation–deintercalation processes in these crystals is analogous to that in GaSe crystals.

Additional investigations of light absorption by GaSe crystals of the first type were performed within the IR range of 2 to 5 μm, where vibrations of water molecules in the condensed state and in the form of individual molecules as well as the carbon dioxide ones are active. Dimensions of the studied crystals were identical to those described earlier. The investigations did not find the bands related with water molecules in the condensed state. At the same time, after subtraction of contribution made by atmosphere water vapors and carbon dioxide gas, there remain weak bands related with molecules of H_2O and CO_2 inside the crystals. Within the error of experiments (2 cm^{-1}), using the CO_2 molecules as an example, one can state that the intercalation of these molecules into the layered matrix results in neither changes in frequencies of "fan" vibrations inherent to these molecules nor the changes in the value of splitting between them.

4.3. Conclusions

The obtained results enabled us to make the following conclusions:

1. The layered crystals are well intercalated with water vapors from Earth atmosphere and deintercalated in the process of evacuation.

2. Water in the crystals of the first type is available in the molecular state close to that in Earth atmosphere.

3. Water molecules are in the state that is not bound with the very crystal.

4. Under long-term storage in ambient atmosphere, the concentration of water molecules in the subsurface area (10–20 mkm) of layered crystals becomes comparable with the concentration of molecules in the very matrix.

5. In the crystals, volume of which is larger than $(1 \times 1 \times 1)$ cm^3, the integrated concentration of water molecules under a long-term storage can exceed the concentration of water vapors in atmosphere by 2 to 3 times.

6. The concentration of H_2O molecules sharply drops when deepening inside the crystal bulk.

7. Decreasing the crystal dimensions down to $(10 \times 10 \times 10)$ μm^3 allows to enhance the rate of intercalation–deintercalation of water molecules in crystals of these dimensions.

5. Low-temperature investigations of photoluminescence spectra taken from InSe and GaSe crystals intercalated with hydrogen and hydrogen-containing molecules: comparative analysis

Photoluminescence spectra of InSe and GaSe crystals nondoped or doped with various impurities have been rather well studied by many authors for approximately 50 years. Therefore, being based on the previous results, one can study the influence of intercalation with hydrogen and hydrogen-containing molecules (HCM) on electron (exciton) excitations in crystalline matrixes of these crystals. Electron transitions in HCM themselves or in their compositions from the ground state to the excited one as well as their following radiative recombination cannot give any contribution to spectra of crystal edge photoluminescence since their transition energies are much higher than the energy of fundamental transitions in these matrixes.

Low-temperature spectra of photoluminescence (PL) were measured at $T = 4.5$ K by using 0.6-m МДР-23 monochromator with the diffraction grating 1200 mm^{-1}. PL excitation was provided with stabilized solid semiconductor lasers of the power 100 mW with the wavelengths 532 and 671 nm. To avoid registration of the second orders from the diffraction grating, we used the edge filters Semrock's LP03-532RS and LP02-830RS. To detect the useful signal, we used the photomultipliers ФЕУ-79 and ФЕУ-62 connected via analog–digital converter with PC. The stabilization of temperature was provided with the accuracy 0.1 K in a liquid helium cryostat by using the system УТРЕКС designed at the Institute of Physics (NAS of Ukraine). The spectral slit width during experiments did not exceed 0.2 meV.

When carrying out this part of the work, we used the samples of InSe and GaSe crystals of the second type of nonintercalated and intercalated with hydrogen, water, alcohol, and toluene in various concentrations.

5.1. GaSe crystals

Shown in Figures 19a to 19f are the PL spectra of GaSe crystals intercalated with toluene, water, and alcohol for 35 and 165 days and measured at $T = 4.5$ K. For comparison, PL spectra of nonintercalated GaSe crystal (Figure 19g) are shown, as well as the crystal doped with Cd (concentration 0.01 mass %) and irradiated with γ-quanta possessing the energies from the range 0 to 34 MeV and doses up to 10_{14} γ-quanta/cm$_2$ (Figure 19h). The latter crystal was earlier investigated in [11]. Adduced in the inserts of Fig. 4.1 (b, d, f) are also the 2D sketches of spatial-coordination ordering inherent to molecules of water, toluene, and alcohol. Figure 19 also illustrates spatial images of these molecules.

As it can be seen in Figure 19, at the temperature $T = 4.5$ K PL spectra of GaSe crystals intercalated with toluene and water, like to the non-doped crystals or those doped with Cd, consist of emission lines related with free excitons (A-line, $\lambda = 587.8$ nm), excitons localized near stacking faults of crystal layers (A_d-lines), and bound excitons (B1 – B6-lines). Besides, one can observe there wide doublet bands P0$^{/}$, P0$^{//}$ that, as shown in [11] using the example of GaSe: 0.01 % Cd crystals, are caused by radiative recombination of carriers from direct (DCB) and indirect (ICD) conduction bands to the acceptor level **a1** located at 70 meV above the top of valence band (VB). In accord with [11], this transition takes place with participation of TO phonons. The energy positions of peaks A, Ad, B1-B6 lines and doublet of the band P0 are summarized in Table 5.

As seen from Figure 19, the intercalation of GaSe crystals with toluene and water for 35 days, if comparing with nonintercalated crystals, results in noticeable changes of PL spectra within the range of free and bound excitons, namely, the intensity of A-line related with emission of

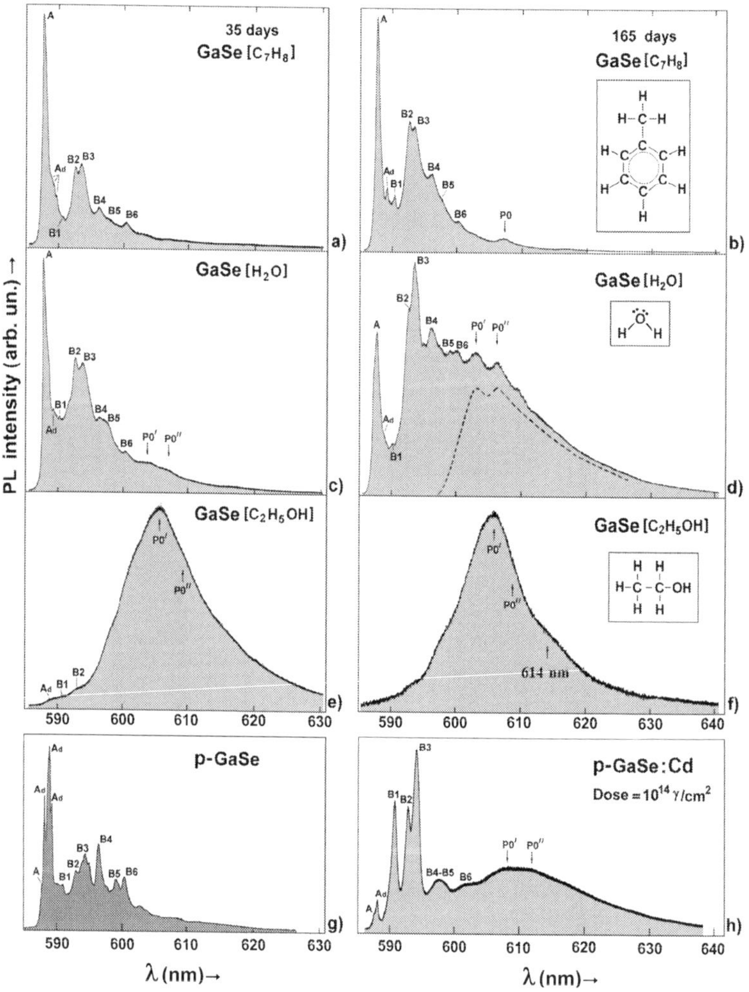

Figure 19. Measured at T = 4.5 K PL spectra of the crystals: a, b—GaSe[C_7H_8]; c, d—GaSe[H_2O]; e, f—GaSe[C_2H_5OH] intercalated for 35 and 165 days, respectively; g—GaSe; and h—GaSe:0.01Cd [11].

free exciton in the ε-GaSe crystal grows, while the intensity of A_d lines related with emission of excitons localized within the ranges of stacking faults in crystal layersops.

It should be noted that the same phenomenon was observed in [11], when doping the crystals with impurities Cd and Zn available in the concentration 0.01 mass % as well as after their "radiation annealing" with γ-quanta of the energy from the range 0 to 34 MeV. In accord with [11], it is related with "curing" the defects by Schottky and by Frenkel, respectively. However, the effect of increasing the intensity of A-line emission and the respective drop in intensity of A_d lines is much better observed in the intercalates GaSe[H_2O] and GaSe[C_7H_8] than in the case of doping with Cd and Zn and the following annealing of these GaSe crystals with γ-quanta.

Crystal	Wave length of emission line and band peaks (nm)								
	A	A_d	B1	B2	B3	B4	B5	B6	PO′, PO″
GaSe	587.8	588.1, 588.6, 589.3	590.9	592.6	594.1	596.5	599.1	600.4	
GaSe[C_7H_8] – 35 days	587.8	588.0, 588.9, 589.7	590.5	592.6	593.7	596.4	597.6	600.5	
GaSe[C_7H_8] – 165 days	587.9	589.2	590.4	592.5	593.5	596.1	597.5	600.2	607.2
GaSe[H_2O] – 35 days	587.9	589.5	590.3	592.1	593.2	596.2	597.4	600.3	603.5, 606.9
GaSe[H_2O] – 165 days	587.9		590.1	592.6	593.7	596.3	599.0	600.1	603.1, 606.3
GaSe[C_2H_5OH] – 35 days	588.9		590.7	593.0					605.8
GaSe[C_2H_5OH] – 165 days									606.0
GaSe:0.01Cd and dose = 10^{14} γ/cm^2) (data from [11])	587.6	588.5		591.5	593.6			600.4	608.0, 612.0

Table 5. Positions of emission peaks for free and bound excitons at $T = 4.5$ K typical for GaSe crystals non-intercalated and intercalated with HCM.

The following growth in the time of intercalation in these solutions for 165 days results in increasing the emission intensity within the range of P0-bands. Despite the fact that this effect is more clearly observed in the intercalates GaSe[H_2O] (see the dashed line in Figure 19d) than in the intercalates GaSe[C_7H_8], the integrated intensity of the emission spectra corresponding to these intercalates (obtained for 165 days) remains the same as in those intercalated for 35 days.

At the same time, attention is attracted by the fact that the shape of PL spectrum typical for the crystal GaSe[H_2O] intercalated for 35 days is practically identical to that of the crystal GaSe[C_7H_8] obtained by intercalation for 165 days. As follows from Table 2 (see Section 3.2), it correlates with measurements of concentrations of H_2O molecules (12.0%) obtained by the intercalation of GaSe for 35 days and of C_7H_6 molecules (9.9%) obtained by intercalation for

165 days. Also, we take into account that the redistribution of the emission intensity, which is observed in the long-wave range of the spectrum, is caused by transitions of carriers from DCB ($P0^{/}$) and ICB ($P0^{//}$) to the **a1** acceptor level with participation of TO phonons. However, as it follows from Figure 19 and Table 6, the energy positions of P0-bands in the case of intercalates $GaSe[H_2O]$ and $GaSe[C_7H_8]$ are shifted into the high-energy range.

The intercalation of GaSe crystals with alcohol (Figures 19c and 19f) results in practically total degradation of emission lines related with free excitons and those localized near stacking faults in crystal layers, as well as excitons bound at point defects. In this case, at $T = 4.5$ K, PL spectra demonstrate only wide weakly structured bands with the peaks at 606.0 ± 0.5 nm ($P0^{/}$, $P0^{//}$) and shoulder close to 614 nm. However, in this case, the integrated intensity of PL spectrum remains practically the same as those in GaSe crystals intercalated with water or toluene. The above energy positions of these bands allow their identification with radiative recombination of carriers from DCB and ICB onto the **a1** acceptor level located near 63 meV above the VB top, which takes place with participation of TO phonons.

According to the data of Table 5, one can show that the energy position of **a1** acceptor level (which for nondoped GaSe crystal is located at 70 meV above the VB top) slows down in the intercalates $GaSe[C_7H_8]$, $GaSe[C_2H_5OH]$ and $GaSe[H_2O]$ by 2.7, 6.7, and 15.8 meV, respectively.

5.2. InSe crystals

Shown in Figures 20a to 20f are low-temperature ($T = 4.5$ K) PL spectra of InSe crystals intercalated with toluene, water, and alcohol for 35 and 165 days. For comparison, shown in Figures 20g and 20h are PL spectra of InSe crystals nondoped and doped with Cd at the concentration 0.01 mass % previously investigated in [19]. The PL spectrum of nondoped InSe crystal (Figure 20g) is superposition of emission lines that correspond to (i) free excitons (A-line, $\lambda = 928.0$ nm), (ii) excitons located near stacking faults in crystal layers (A_d lines), (iii) excitons localized in the density-of-state tails (A_l lines), and (iv) excitons related with point defects of crystalline structure (B1–B6 lines), which are related with defects by Frenkel and Schottky (see [19]).

Contrary to ε-GaSe, InSe crystals have rhombohedral symmetry of the crystalline lattice (γ-modification), which is lower than that of ε-GaSe [2]. Therefore, the partial disordering of crystal layers in InSe results in appearance of stacking faults in layers, namely, the areas of mixed ε- and γ- modifications. In these areas, the band gap is larger than in the areas of pure γ-modification, and excitons located in the mixed areas possess higher emission energies. The energy positions of the respective lines and P0-bands at $T = 4.5$ K are summarized in Table 6.

The intercalation of InSe crystals with molecules of toluene, water, and alcohol, as well as the intercalation time, has an essential influence on their optical properties. It is clearly pronounced (Figures 20a to 20f) in an essential decrease in the emission line intensity of bound excitons (B1–B6 lines) and in the respective growth of emission lines related with free excitons as well as excitons bound with stacking faults in layers and in the density-of-state tails. For comparison, it should be noted that the same changes take place (Figure 20h) in PL spectra of InSe crystals doped with Cd [19] or electrochemically intercalated with hydrogen (see insert in

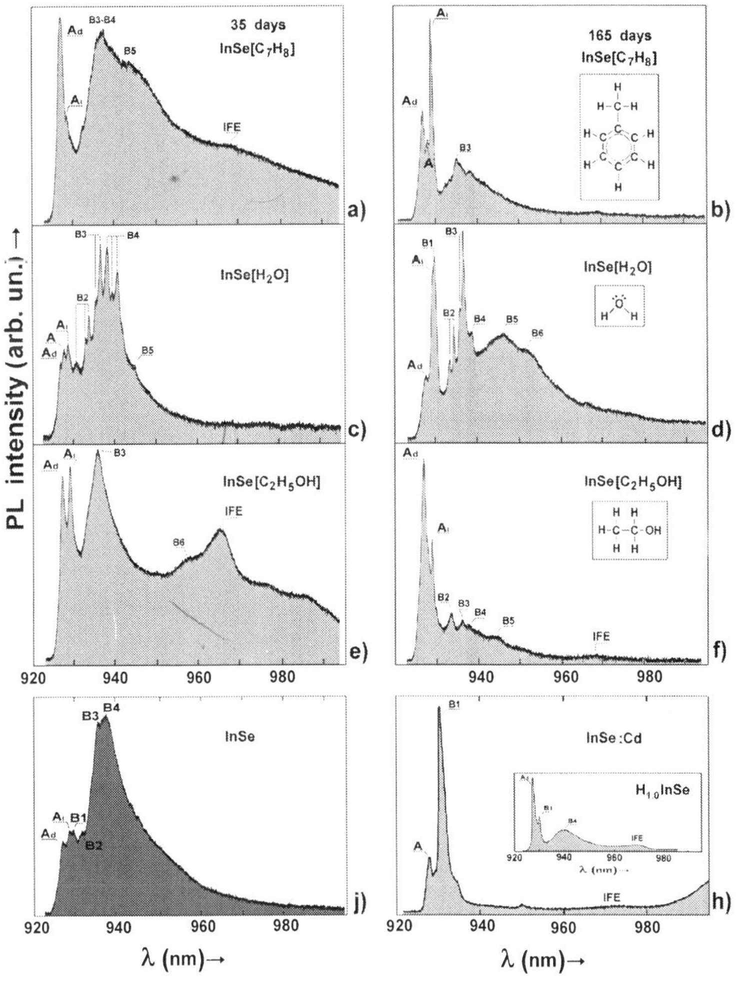

Figure 20. PL spectra of the crystals at the temperature T = 4.5 K: a, b—InSe[C_7H_8]; c, d—InSe[H_2O]; e, f—InSe[C_2H_5OH], intercalated for 35 and 165 days, respectively; g—InSe; and h—InSe:Cd [19]. Insert in h—$H_{1.0}$InSe.

Figure 20h). Thereof, contrary to GaSe crystals, intercalation with hydrogen or hydrogen-containing molecules, like to doping with elements of the second group of periodic table, results in essential improvement of the crystalline lattice in InSe crystals.

An increase in the time of intercalating the InSe crystals with toluene, water, or alcohol leads to improvement of PL spectra: A_d lines begin to dominate, and the emission line of free exciton typical for γ-modification appears in crystals intercalated with toluene. The value of blue shift observed for A_d lines in the intercalated crystals relatively to A-line in nonintercalated γ-InSe reaches the value 1.5 meV. It is also typical that B2–B4 lines corresponding to emission of bound excitons in the crystals intercalated with water are split. In this case, the split states are located in the long-wave range relatively standard positions of these lines, and the value of maximum splitting for these lines approaches to 2.0 meV. It is indicative of a rather complex structure of defects that provide bonding these excitons.

Crystal	Wave length of emission line and band peaks (nm)									
	A_d	A	A_l	B1	B2	B3	B4	B5	B6	IFE
InSe	927.5		929.0	930.0	932.0	936.0	938.0	-	-	-
$InSe[C_7H_8]$ – 35 days	927.0		928.5							968.0
$InSe[C_7H_8]$ – 165 days	927.0	928.0	928.9			935.0				
$InSe[H_2O]$ – 35 days	926.9	927.9	929.1	-	931.9 933.1 933.9	935.6 936.8	938.5 939.7 941.0	945.0	-	-
$InSe[H_2O]$ – 165 days	927.6		9292.4	929.8	933.3 934.3	935.9 936.5	938.7	946.1	951.9	-
$InSe[C_2H_5OH]$ – 35 days	927.3		929.1			936.0			957.4	965.3
$InSe[C_2H_5OH]$ – 165 days	926.8		928.9		933.5	936.1	937.6	944.2		968.0
InSe:0.01Cd (data [19])		927.9	-	930.2	-	935.5	-	-	950.0	972.0

Table 6. Positions of PL lines for free and bound excitons at $T = 4.5$ K in nonintercalated and intercalated InSe crystals

Besides, in crystals InSe intercalated with alcohol for not so long time (35 days), one can observe the wide band peaking at 966 nm that is caused by radiative recombination of free excitons from indirect conduction band with participation of TO phonons. With increasing the time of intercalation, this band decays.

5.3. Discussion

The performed investigations of PL spectra of InSe and GaSe crystals intercalated with toluene, water, and alcohol allows stating that intercalation with HCM that contain carboxyl CH and hydroxyl OH groups exert different influence on optical properties of these crystals. In InSe crystals, intercalation with HCM considerably decreases the intensity of emission lines related with bound excitons, which is a direct confirmation of improvement of the crystalline lattice as a consequence of decreasing the amount of defects responsible for exciton bonding.

Vice versa, in GaSe crystals, increasing the time of intercalation with water and toluene results in enhancing the intensity of radiation from bound excitons, while the structure of the line inherent to free excitons is not practically improved. At the same time, the intercalation of GaSe crystals with alcohol leads to full disappearance of emission lines both of free and bound excitons, which is quite opposite to that observed in InSe crystals. In addition, the intercalation of GaSe crystals with HCM results in lowering the energy position of the **a1** acceptor level relatively to the VB top.

We consider that these differences are caused by peculiarities of covalent bonds inside crystal layers of these crystals. Thus, InSe crystals are more ionic than the GaSe ones [20]. It defines the formation of a great number of various defects, which results in appearance of different type conductivity, namely, *n*-type in InSe and *p*-type in GaSe. To expand discussion, let us consider results of the studies [11, 19], where it was shown that doping the crystals InSe and GaSe with low concentration of impurities (0.01 mass % of Cd or Zn) and the following irradiation with γ-quanta results in the growth of integrated intensity of radiative recombination. This growth takes place on the background of considerable lowering the PL intensity of bound excitons relatively to that of free excitons. In this case (see Figure 20), the PL spectrum of the InSe crystal doped with Cd impurity approaches by its shape and structure to the spectrum of InSe intercalated with hydrogen, toluene, or alcohol.

Vice versa (see [11, 19]), doping the crystals InSe and GaSe with Ge or Sn results in a sharp decay of the integrated intensity of radiative recombination. It allows making an assumption that doping with elements of the second and fourth groups of periodic table influences the growth of these crystals in a different manner. Moreover, even at low concentrations (0.01 mass %), impurities of Cd, Zn, Ge, or Sn are nonuniformly distributed in these crystals, and their concentration can differ almost by one order from one point to another.

At the same time (see [19]), when doping the InSe crystals with Cd or Zn, the type of crystal conductivity changes from *n*- to *p*-type, and in the spectral range after 1.0 mkm (Figure 20h), one can observe the P1–P3 emission bands related with carrier transitions from DCB and ICB to the deep acceptor level as well as transitions between donor and acceptor levels with participation of optical TO phonons. The similar wide bands are also observed in GaSe crystals of *p*-type conductivity after their doping with low concentrations of Cd or Zn impurities [11, 19].

However, contrary to doping the InSe crystals with Cd or Zn, their intercalation with hydrogen or HCM leads neither to changing their type of conductivity nor to appearance of new additional wide bands in the range after 1.0 mkm.

Also in [1], the authors considered the case of passivating the defects with hydrogen atoms in InSe and GaSe crystals of *n*- and *p*-type. According to this work, under electrochemical intercalation of InSe and GaSe crystals with hydrogen, the latter is embedded into interlayer and intralayer spaces and recombines there up to the molecular state. At the same time, a small part of hydrogen remains in the atomic state and passivates point defects in the crystal. When passivating the donor (see Figures 21a and 21b), proton is localized closer to the defect, while passivating the acceptor, proton is more movable. It is confirmed by the fact that in EPR spectra g-factor and the spectral width in H_XInSe intercalates are considerably lower than in H_XGaSe ones [1].

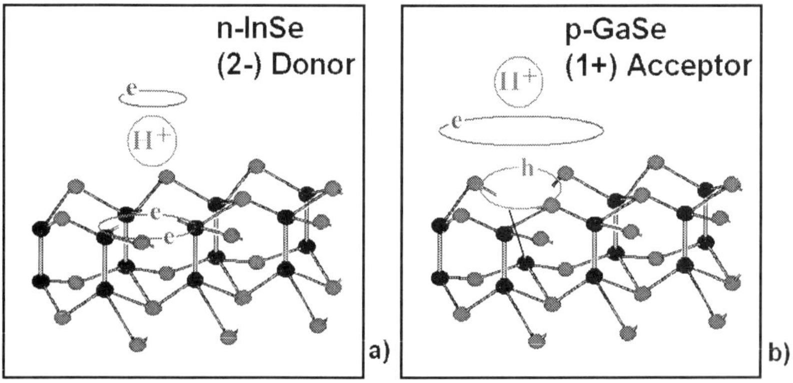

Figure 21. a, b – the simplest defects of donor and acceptor type in *n*-InSe and *p*-GaSe crystals and the scheme of their passivation with hydrogen.

In accord with [21], passivation with atomic hydrogen also results in lowering the activation energy of donor and acceptor levels, since in H_XInSe intercalates at low temperatures ($T = 70$ K) one can observe the growth by one order in the free electron concentration, and the further growth of temperature results in exhaustion of the shallow donor level. The additional optical investigations performed by us also show (see the insert to Figure 20h) that intercalation with hydrogen improves PL spectra of *n*-InSe crystals and does not influence on PL spectra of *p*-GaSe crystals.

At the same time, when intercalating the crystals with hydrogen, the lattice parameter and width of the van der Waals gap grow due to embedding the hydrogen molecules into interlayer and intralayer spaces [1]. The authors of the work [22] investigating the influence of hydrostatic pressure on lattice parameters in InSe crystals showed that increasing the pressure from 10^5 up to 8×10^9 Pa results in essential narrowing the van der Waals gap as a consequence of decreasing the distance $C_{Se\text{-}Se}$ between the two nearest Se atoms located in the adjacent layers from 0.38 down to 0.33 nm (13%). Inside crystal layers, the distance $C_{In\text{-}In}$ between neighboring In atoms is decreased from 0.279 down to 0.269 nm (3.58%), and the distance $C_{In\text{-}Se}$ between In and Se atoms—from 2.65 down to 2.59 nm (2.26%). However, as a consequence of the fact that the angle φ between atoms Se–In–In grows from 119.3° up to 121.3° (1.5%), the thickness of crystal layer decreases only by 0.26%—from 0.536 down to 0.5345 nm.

The extrapolation of data from the studies [1, 22] as well as comparison of lattice parameters of InSe [6, 22] and GaSe [5] with sizes of molecules allows to conclude that the natural intercalation of these crystals with molecules of water, alcohol, and toluene results in their embedding into the interlayer space, which causes growth of the van der Waals gap in respective places. In this case, electrochemical intercalation leads to physical sorption of hydrogen in crystals (since hydrogen can be deintercalated from InSe and GaSe crystals under usual evacuation at the temperatures 110–120°C with participation of full-symmetric half-layer A-phonons able to change the thickness of crystal layers and the intralayer space [1]), while in the case of natural intercalation with water, alcohol, and toluene there takes place chemical sorption of molecules in the interlayer space.

To check up general properties and differences that occur in InSe and GaSe crystals after doping with Cd, Zn, Ge, or Sn atoms, or after intercalation with water, alcohol, and toluene, let us consider the simplest defects typical for these crystals (Figure 21).

In accord with Figure 21a, the absence of Se atom inside the layer makes it impossible to allocate 2/3 portion of electron charge to each of the nearest atoms for the creation of the natural sp^3-hybrid bond. It is the way with which the simplest (2–) donor arises in *n*-InSe crystals. Simultaneously, the absence of Ga atom in *p*-GaSe crystal (Figure 21b) results in the creation of the (1+) acceptor state since unpaired *s*-electron from the nearest Ga atom provides the hybrid bond with three Se atoms of the upper monolayer.

In the course of doping the *p*-GaSe crystals with impurities Cd or Zn (Figure 21b), it becomes possible for Ga vacancies to be substituted with Cd or Zn atoms with the simultaneous creation of a new (1+) acceptor as a consequence of deficit of one valence electron in Cd or Zn atom. Then there arises a set of absorption and emission bands in the long-wave spectral range, which is caused by transitions of carriers from direct and indirect conduction bands on the deep acceptor level, or by transitions between donor and acceptor levels in *p*-GaSe:(Cd, Zn) crystals.

Doping the InSe crystals with impurities, Cd or Zn changes their type of conductivity. It becomes possible (see Figure 21a) when three upper In atoms are simultaneously substituted with Cd or Zn atoms. This more complex defect becomes the (1+) acceptor. The concentration of impurities near these defects should be higher since it changes the type of conductivity in InSe crystals, which by analogy with the *p*-GaSe:(Cd, Zn) crystals results in appearance of wide bands in PL spectra of *p*-InSe:(Cd, Zn) that are caused by radiative recombination of carriers from DCB and ICB onto the deep acceptor level. Moreover, the availability of another type defects enables to explain the reason for nonuniform distribution of impurities in these crystals.

In the case of crystals intercalated with HCM, embedded molecules, except hydrogen, contain either carbon atoms (toluene, alcohol) or hydroxyl group (alcohol, water). The investigations performed by us showed that the intercalation of *p*-GaSe and *n*-InSe crystals with toluene leads to changes in PL spectra, which are similar to those observed after intercalation with hydrogen. It allows concluding that passivation of defects with toluene is identical to that with hydrogen. However, in the case of toluene molecules, hydrogen ions are symmetrically located relatively to the main axis of the molecule and possess other electro-negativity than those in the methyl tail. It leads to different positions of toluene molecules in the van der Waals gap of *n*-InSe and *p*-GaSe crystals that contain different defects.

Thus, in *n*-InSe crystals intercalated with toluene and hydrogen one can observe "curing" the defects since emission lines of excitons localized on various defects in crystals disappear. By contrast, in the case of *p*-GaSe growth of the intercalation time "worsens" crystal quality—there appear P0-bands, and the emission intensity of bound excitons increases. It can be possible, when a toluene molecule is differently oriented relatively to (2–) donor and (1+) acceptor. As the (2–) donor has higher electro-negativity, the toluene molecule approaches to this defect with its massive benzene ring and becomes more stabilized in space. In the case of (1+) acceptor, the molecule approaches to the defect with its tail C-H_3. In this situation, there can take place deviation of the toluene molecule near this defect, which results in decay and binding the excitons as well as to enhancing radiative recombination via the defect centers of this crystal.

Unlike to toluene molecules, those of water and alcohol contain the hydroxyl group OH. In *p*-GaSe crystals, it results in essential growth of the P0-band in PL spectra that is caused by radiative recombination of carriers from DCB and ICB onto the acceptor level a1 with emission of TO phonons. Besides, the intensity of exciton emission is considerably lowered, since water and alcohol molecules, being near the (+1) acceptor, essentially disturb the lattice of *p*-GaSe crystal.

Contrary to *p*-GaSe crystals, in *n*-InSe crystals intercalated with water and alcohol (like those with embedded hydrogen and toluene) curing the point defects takes place. However, in this case (2–) donor is passivated with the hydroxyl group. As the water molecule is more symmetric than the alcohol one, and its size practically coincides with that of defect, it can vary between several symmetric positions relatively to this defect. In the crystal lattice, it results in splitting the exciton state related with this defect.

5.4. Conclusions

The performed investigations of low-temperature (T = 4.5 K) PL spectra of GaSe and InSe crystals intercalated with hydrogen-containing molecules (HCM) of water and alcohol, their comparison with the spectra of GaSe and InSe crystals intercalated with hydrogen or doped with impurities Cd, Zn, Ge, Sn [11, 19], as well as review of the simplest defects carried out in discussion for the crystals of *n*- and *p*-type enabled to explain differences of optical properties inherent to *p*-GaSe and *n*-InSe crystals and to draw the following conclusions:

i. HCM are predominantly located in the interlayer space and passivate point defects of *p*-GaSe and *n*-InSe crystals.

ii. The availability of two conductivity types, namely, *p*-type in GaSe and *n*-type in InSe, leads to different influence on their optical properties.

In particular, in *p*-GaSe crystals intercalation with HCM possessing the OH group results in considerable disturbance of the crystalline lattice around the (+1) acceptor, as a consequence of which emission lines inherent to free and bound excitons disappear, while radiative recombination of carriers from DCB and ICB onto the a1 acceptor level located at 70 meV above the VB top grows. This effect of increasing the intensity of radiative recombination of carriers from DCB and ICB onto the a1 acceptor level in *p*-GaSe is observed with increasing the intercalation time as well as when transferring along the direction toluene → water → alcohol.

When transferring from nonintercalated GaSe to the HCM-intercalated ones along the set GaSe → GaSe[C_7H_8] → GaSe[C_2H_5OH] → GaSe[H_2O], the energy position of the a1 acceptor level in the forbidden gap is lowered in accord with the sequence 70 → 67 → 63 → 56 meV. By analogy with the *p*-GaSe and *n*-InSe crystals intercalated with hydrogen, it has been made an assumption that the observed lowering the position of activation energy for the a1 acceptor level is caused by passivation of this defect with HCM.

By contrast, in *n*-InSe crystals intercalated with HCM, one can observe curing the point defects, predominantly (2–) donors, which results in the growing intensity of emission lines related with free excitons and practically full disappearance of lines inherent to bound excitons as well as emission bands of indirect free excitons with participation of optical phonons. In this case, the effects of curing the defects in *n*-InSe are more clearly pronounced after intercalation with HCM containing the molecules of carboxyl group (CH) or molecular hydrogen than in those with the hydroxyl group (OH)–water or alcohol.

Also, after the intercalation of *n*-InSe crystals with water, splitting the emission lines of bound excitons is observed, which is caused by splitting the exciton states related with the (2–) donors. However, intercalation with HCM does not practically change the integrated intensity of radiative recombination for excitons and free carriers in *p*-GaSe and *n*-InSe crystals.

6. Summary

In this review, in the wide energy range from energies of electron transitions between inner shells of intercalated atoms and those typical for energies of electrons providing valence bonds in the crystalline matrix down to the energies of intrinsic vibrations of atoms (nuclei) of molecules intercalated into the matrix, we have presented the results of investigations by using electron-microscopic, energy-dispersive, and optical methods. These methods allow to study spectra within the range of fundamental electron transitions in the matrix as well as in the range of IR vibrations of intercalated molecules in layered InSe and GaSe crystals with embedded molecular hydrogen or hydrogen-containing molecules (HCM) of water, alcohol, and toluene. Described also have been the methods for the intercalation and deintercalation of HCM.

It has been shown that these crystals in the course of intercalation free take (and during deintercalation release) not only atomic hydrogen but also HCM. This process takes a look of a hysteresis loop—it is retarded during intercalation and is faster during deintercalation when heating and evacuating in a vacuum chamber. Adduced have been the empirical equations relating these two processes and their temporal estimates. We have also determined the optimal dimensions of the crystals (including their powders) to accelerate the process of intercalation–deintercalation.

As InSe and GaSe crystals are characterized by two different types of conductivity, it results, when intercalating with hydrogen and HCM, in predominant recombination of free and bound excitons in *n*-InSe crystals, while in *p*-GaSe crystals processes of free carrier recombination via impurity levels are predominant. These processes have no effect on the integrated intensity of emission from the crystal within the range of fundamental transitions, since the energy of valence electron transitions for HCM lies in the UV range.

The results obtained in this work with account of the conclusions made by us in [1] are indicative of promising applications of these layered matrixes (and powders from them) when creating operation elements of solid hydrogen accumulators. At the same time, these layered InSe and GaSe crystals themselves with different types of conductivity and their thin-layer intercalates as well as other similar compounds can be considered as promising candidates for the creation of various hybrid layered structures for flat flexible displays. The obtained in this work results for various cases of volume dimensions, conditions and terms of their operation can be used for solving these tasks, too.

Acknowledgements

Performing these investigations became possible owing to the project no. 0108U000252 within the framework of the departmental subject area in the National Academy of Sciences of Ukraine in accord with the Program for Scientific Instrument Engineering.

Author details

Yuriy Zhirko[1*], Vasiliy Grekhov[1], Nikolay Skubenko[1], Zakhar Kovalyuk[2] and Taras Feshak[2]

*Address all correspondence to: zhirko@nas.gov.ua

1 Institute of Physics National Academy of Sciences of Ukraine, Kiev,, Ukraine

2 Chernivtsy Department of the Institute for Material Science Problems National Academy of Sciences of Ukraine, Chernivtsy, Ukraine

References

[1] Zhirko Y, Trachevsky V, Kovalyuk Z. Some physico-chemical aspects of layered crystal application for solid state hydrogen storage. InSe and GaSe crystals. In: Jianjun Liu editor. Hydrogen storage. InTech, 2012, Chapter 9; p. 211–242. (ISBN) 978-953-51-0731-6

[2] Belen'kii G, Stopachinskii V. Electronic and vibrational spectra of 3–6 layered semiconductors. Sov. Phys. Usp. 1983; 26: 497–517.

[3] Polian A, Kunc K, Kuhn A. Low-frequency lattice vibrations of ∂-GaSe compared to ε- and γ-polytypes. Solid State Commun. 1976; 9: 1079–1082.

[4] Ghosh P.H. Vibrational spectra of layer crystals. Appl. Spectrosc. Rev. 1983; 19: 259–323.

[5] Kuhn A, Chevy A, Chevalier R. Crystal structure and interatomic distance in GaSe. Phys. Status Solidi B. 1975; 31: 469–473.

[6] Goni A, Cantarero A, Schwarz U, Syassen K, Chevy A. Low-temperature exciton absorption in InSe under pressure. Phys Rev B. 1992; 45: 4221–4226.

[7] Martines-Pastor, Segura A, Valdes J. Electrical and photovoltaic properties of indium-tin-oxide/p-InSe/Au solar cells. J. Appl. Phys. 1987; 62: 1477–1483.

[8] Lebedev A, Rud' V, Rud' Y. Photosensitivity of geterostructures porous silicon-layered $A^{III}B^{VI}$ semiconductors. Fizika i Tekhnika Poluprovodnikov (in Russian). 1998; 32: 353–355.

[9] Shigetomi S, Ikari T. Electrical and photovoltaic properties of Cu-doped *p*-GaSe/*n*-InSe heterojunction. J. Appl. Phys. 2000; 88: 1520–1524.

[10] Grigorchak I, Zaslonkin A, Kovalyuk Z, et al. Patent of Ukraine, 2002, Bulletin No. 5, No46137.

[11] Zhirko Y, Skubenko N, Dubinko V, et al. Influence of impurity doping and γ-irradiation on the optical properties of layered GaSe crystals. J. Mater. Sci. Eng. B. 2012; 2: 91–102.

[12] Kozmik I, Kovalyuk Z, Grigorchak I, Bakchmatyuk B. Preparation and properties of hydrogen intercalated gallium and indium monoselenides. Isvestija AN SSSR Inorganic materials (in Russian). 1987; 23: 754–757.

[13] Zhirko Y, Kovalyuk Z, Pyrlja M, Boledzyuk V. Application of layered InSe and GaSe crystals and powders for solid state hydrogen storage. In: Vezirogly N et al editors. Hydrogen Materials Science and Chemistry of Carbon Nano-materials. Springer ©; 2007 p.325–340.

[14] Zhirko Y, Kovalyuk Z, Klad'ko V. et al. Investigation of hydrogen intercalation in layered crystals InSe and GaSe. In: Proceedings of (HTM-2007) 15 International Conference Hydrogen Economy and Hydrogen Treatment of Materials; 21–25 May 2007; Donetsk, Ukraine. 2007. 2, p. 606–610.

[15] Zhirko Y, Kovalyuk Z, Pyrlja M, Boledzyuk V. Optical investigation of hydrogen intercalation–deintercalation processes in layered semiconductor γ-InSe crystals. In: Hydrogen Materials Science and Chemistry of Carbon Nanomaterials. NATO Science Series II: Mathematics, Physics and Chemistry. 2004; 172. p. 519–530.

[16] Demchina L, Kovalyuk Z, Mintjanskii I. Ommic contacts preparation for A3B6 layered crystals. Pribory I Tekhnika Eksperimenta. 1980; 2: 219–221.

[17] Grigorchak I, Kovalyuk Z, Yurtzenjuk S. Preparation and properties of intercalated $A^{III}B^{VI}$ layered junctions. Izveseija AN SSSR. Ser. Neorg. Mater. 1981; 17: 412–415.

[18] Zhirko Y. Role of intrinsic defects in the process of synthesizing the layered crystals GaSe. In: Proceedings of 14-th European Conference of Solid State Chemistry (ECSSC14). 7–10 July 2013; Bordeaux. France. p.139

[19] Zhirko Y, Skubenko N, Dubinko V, Kovalyuk Z, Sydor O. Influence of γ-Irradiation on the Optical Properties of Pure or Impure Zn, Cd and Ge-Doped Layered InSe Crystals. J. Mater. Sci. Eng. B. 2013; 3:162–174.

[20] Nakanishi F, Matsubara T. Note on ionisity of kayered conhounds GaS, GaSe, and InSe. J. Phys. Soc. Japan. 1982; 516: 1339–1340.

[21] Zhirko Y, Kovalyuk Z, Zaslonkin A, Boledzyuk V. Photo and electric properties of hydrogen intercalated InSe and GaSe layered Crystals. In: Proceedings of Int. Confer. Nano/Moleclar Photochemistry and Nanomaterials for Green Energy Development (Solar'10). 15–17 February 2010; Cairo, Egypt; 2010. p. 48–49.

[22] Olguin D, Cantarero A, Ulrich C, Suassen K. Effect of pressure on structural properties and energy band gaps of γ-InSe. Phys. Status Solidi B. 2003; 235: 456–463.

Section 3

Hydrogen Storage

Chapter 3

Metal Amidoboranes and Their Derivatives for Hydrogen Storage

Hailiang Chu, Shujun Qiu, Lixian Sun and Fen Xu

Additional information is available at the end of the chapter

http://dx.doi.org/10.5772/61048

Abstract

As potential hydrogen storage mediums, ammonia borane and its derivatives have been paid an increasing attention owing to their higher hydrogen capacities and facile dehydrogenation properties under moderate conditions. In this chapter, we presented extensive studies on thermodynamic tailoring of dehydrogenation of metal amidoboranes, metal borohydride-ammonia borane complexes, and metal amidoborane ammoniates as well as their derivatives, with special focus on the syntheses, crystal structures, and dehydrogenation properties. Finally, future perspective was given toward the practical applications.

Keywords: Hydrogen storage, dehydrogenation, metal amidoborane, B-N-containing compounds, thermodynamics, kinetics

1. Introduction

Mainly driven by two global issues of energy crisis and environmental pollution, great attention to a relatively new concept of the hydrogen economy has been paid in the last decades. However, hydrogen storage is one of bottlenecks in hydrogen economy. Generally, the materials with high hydrogen content have received an increasing interest for chemical storage concepts. As a chemical hydride, ammonia borane (NH_3BH_3, AB) has attracted significant attentions owing to the extremely high hydrogen capacity up to gravimetric density of 19.6 wt% and volumetric density of 146 g H_2 L^{-1} [1]. Unlike metal borohydrides, solid crystalline AB is not very sensitive to air at ambient condition. In AB molecules, there are both hydridic H atoms in BH_3 groups and protic H atoms in NH_3 groups. Consequently, the chemical potential of the interaction between protic $H^{\delta+}$ and hydridic $H^{\delta-}$ to the formation of hydrogen has been known as one of the prime driving powers for hydrogen release [2,3].

However, the relatively high kinetic barriers are observed in its multistep decomposition, which heightens the temperatures for hydrogen release more than 100 °C [4,5]. Other disadvantages of AB dehydrogenation consist of the discharge of some detrimental by-products (borazine and ammonia) and severe sample expansion and foaming. Moreover, the intrinsic exothermicity of dehydrogenation reveals that reversible hydrogenation under moderate conditions is thermodynamically impracticable. In recent years, a number of techniques including nanoconfinement [6-8], using ionic liquid medium [9], and catalytic modifications [10-14] were successfully employed in extensive studies. More importantly, another strategy proposed to improve the properties of AB thermolysis is the formation of metal-substituted AB-compounds, i.e., metal amidoboranes (MABs), which have been proven to be a practical method for tuning the dehydrogenation thermodynamics of AB [15-20]. Because of the quite high hydrogen content and reasonable dehydrogenation properties, this new type of materials has been paid intensive attention and grown to become one of the likeliest candidates for storing hydrogen.

In this chapter, we briefly summarize the recent progress of nanoconfinement and nanocatalysis of AB, metal amidoboranes (including single- and mixed-metal amidoboranes), and their derivatives (including metal amidoborane ammoniates, metal borohydride-ammonia borane complexes, and other complexes related to AB or MAB), with a particular emphasis on the synthesis, crystal structure, and dehydrogenation.

2. Nanoconfinement of AB

The pioneering work on AB nanoconfined in porous materials was performed by Autrey and his coworkers [6]. They loaded AB into SBA-15 (a kind of mesoporous material with specific surface area of about 900 $m^2\ g^{-1}$, pore size of 7.5 nm in diameter, and porous volume of 1.2 $cm^3\ g^{-1}$) by impregnating SBA-15 in a saturated methanol solution of AB. The obtained sample was determined to contain 50 wt% AB (denoted as AB:SBA-15). Compared to pristine AB, the sample nanoconfined in SBA-15 has two significant differences, as shown in Figure 1. First, the dehydrogenation peaks shifted to lower temperatures. Second, the great suppression of borazine evolution was successfully achieved, which indicates that the dehydrogenation reaction is probably tuned. The change of reaction pathway was further verified by differential scanning calorimetry (DSC) and nuclear magnetic resonance (NMR) experiments. From isothermal DSC test, the dehydrogenation reaction enthalpy of AB:SBA-15 was found to significantly reduce to 1 ± 1 kJ $(mol\ AB)^{-1}$, compared to pristine AB of 21 ± 1 kJ $(mol\ AB)^{-1}$. Also, NMR tests showed that the post-dehydrogenated product was only polyaminoborane (PAB) for AB:SBA-15 upon heating to 85 °C, which is different from pristine AB.

Isothermal dehydrogenation showed that the releasing rate of hydrogen from AB:SBA-15 was much more rapid than that from pristine AB. As an example, it needed less than 10 min for AB:SBA-15 to reach 50% reaction conversion at 80 °C, while 290 min for pristine AB. Also, a much lower activation energy of 67 ± 5 kJ mol^{-1} was determined for dehydrogenation from AB:SBA-15. However, the total amount of released hydrogen was undoubtedly sacrificed after AB loading into SBA-15, which was not specifically mentioned in their work.

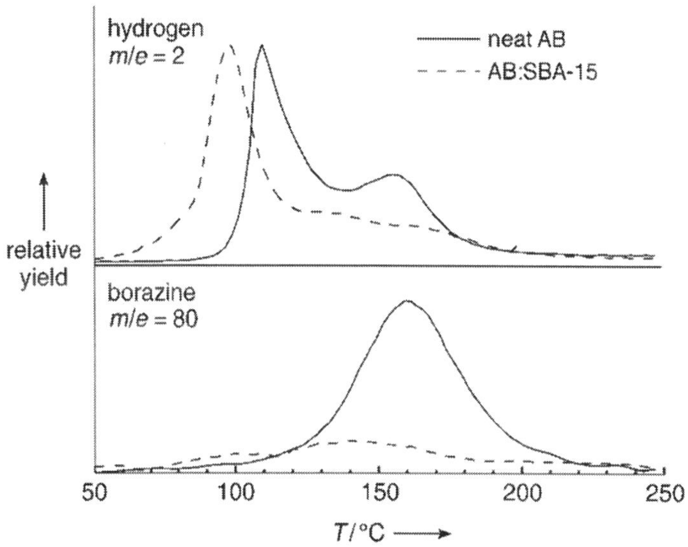

Figure 1. TPD/MS tests of AB nanoconfined in SBA-15 (AB:SBA-15). For comparison, neat AB was also included. Reproduced from Ref. [6] with permission.

In another study [7], carbon cryogel with specific surface area of 300 m^2 g^{-1}, pore diameter in the range of 2-20 nm, and pore volume of 0.7 cm^3 g^{-1} was selected as a nanoscaffold to incorporate AB. Sample with a loading of 24 wt% AB into carbon cryogel was obtained (denoted as C–AB nanocomposite) by wet impregnation using tetrahydrofuran (THF) as solvent. Nitrogen physisorption test revealed that there is a loss of ~30% in pore volume after loading AB. From Figure 2, it was found that carbon cryogel as a nanoscaffold exhibited an even more pronounced effect on dehydrogenation than SBA-15. The peak for hydrogen release shifted to a low temperature of ~90 °C, and no signal for borazine was observed throughout the experiment. However, the total amount of released hydrogen was significantly less than that from pristine AB (9 wt% *vs.* 13 wt%). When confined in carbon cryogel, AB underwent the expected reactions for dehydrogenation. Unexpectedly, the enthalpy of dehydrogenation reaction for C–AB nanocomposite was determined to be an extremely high value of ~120 kJ mol^{-1} (compared to pristine AB of 21 ± 1 kJ mol^{-1} and AB:SBA-15 of 1 ± 1 kJ mol^{-1}). However, it remains so far unclear what caused the tremendous increase in reaction enthalpy. The authors suggested that the improvements (for AB loaded into both carbon cryogel and SBA-15) were ascribed to the increase of defect sites and surface energy of AB, resulting from nano-confinement and/or the possible catalysis by the terminal SiO–H groups in SBA-15 and carboxylic groups in carbon cryogel. For carbon cryogel used as a nanoscaffold, the smaller the pore diameter, the lower the temperature for the dehydrogenation. In-depth studies of pore size effect on the structure and dehydrogenation are needed in the forthcoming research.

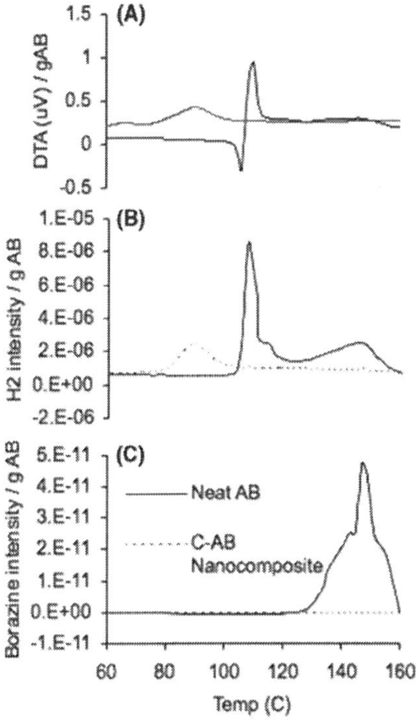

Figure 2. (A) DTA, MS for (B) H_2 and (C) borazine of AB nanoconfined in carbon cryogel (C-AB nanocomposite) at 1 °C/min. For comparison, neat AB was also included. Reproduced from Ref. [7] with permission.

As a newly developed kind of porous material, metal-organic frameworks (MOFs) consist of polydentate organic ligands and metal ions, achieving 3D extended frameworks [21,22]. A unique feature of MOFs is its adjustable porosity, which enables it with special properties like the combination of active metal sites and nanoporosity. Taking advantage of its porosity, an MOF-confined AB system was successfully synthesized by an infusion method using a solvent of methanol [23]. $Y(BTC)(H_2O){\cdot}DMF$ (denoted as JUC-32-Y) was used as a nanoscaffold because of its high specific surface area, suitable pore size, and high thermal stability. Most importantly, it has many unsaturated metal sites after the removal of the terminal H_2O upon heating at 300 °C under vacuum (Figure 3A). After loading AB in 1:1 molar ratio, the framework of JUC-32-Y retained its integrity, even after dehydrogenation of AB/JUC-32-Y. Temperature programmed desorption coupled with mass spectrometry (TPD-MS) results shown in Figure 3B demonstrated that the dehydrogenation of AB/JUC-32-Y started at a lower temperature of ~50 °C, with a peak at 84 °C for most rapid releasing rate, which has a reduction of 30 °C in comparison with pristine AB. Furthermore, no volatile products such as B_2H_6, NH_3, and borazine were detected during the decomposition process. The suppression of ammonia was ascribed to the existence of unsaturated coordination of Y^{3+}. Figure 3C indicates a notable improvement of the dehydrogenation kinetics of AB/JUC-32-Y. At 95 °C, 10.2 wt% H_2 could be released in initial 10 min while 8.0 wt% H_2 at a reduced temperature of 85 °C. Pristine AB displayed no emission of any hydrogen under same conditions. Note that AB/JUC-32-Y could release ~13 wt% H_2 at 95 °C within 3 h, corresponding to 2 Eq of H_2 in pristine AB.

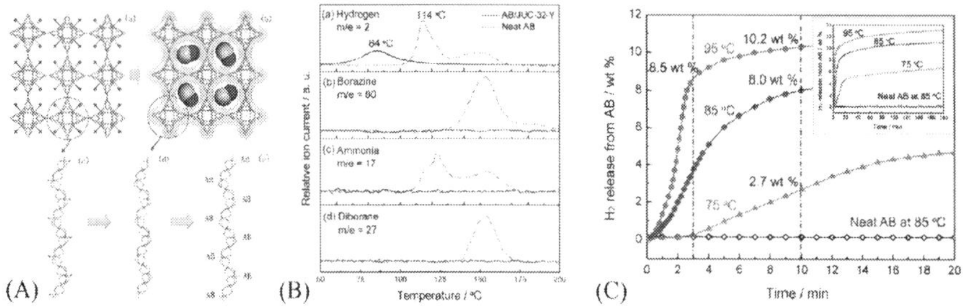

Figure 3. (A) 3D images of (a) JUC-32-Y and (b) AB/JUC-32-Y. Views of 1D chains of JUC-32-Y (c) before and (d) after removal of terminal H_2O and (e) after interaction with AB. Y: bright blue; C: gray; O: pink; N: blue; and B: purple. (B) TPD-MS and (C) Isothermal dehydrogenation of AB/JUC-32-Y. For comparison, neat AB was also included. Reproduced from Ref. [23] with permission.

Soon afterward, Gadipelli et al. [24] reported the confinement of AB in Mg-MOF-74 with 1D pores (denoted as AB/Mg-MOF-74), in which the dehydrogenation kinetics had a remarkable improvement under 100 °C. Furthermore, in the released hydrogen, the detrimental concomitants such as ammonia, borazine, and diborane were not detectable throughout the dehydrogenation process. More importantly, owing to the relatively high specific surface area of 1100 m^2 g^{-1} and low atomic weight of metal cation, Mg-MOF-74 could contain about 26 wt% AB determined by theoretically structural optimization, while only about 8 wt% AB in JUC-32-Y [23]. Almost 14 wt% of pure H_2 could be released at a temperature of 125 °C in one hour for the AB/Mg-MOF-74 sample, which exceeded 13 wt% of H_2 (2 Eq) from pristine AB. Further systematic investigation showed that the dehydrogenation kinetics of AB/Mg-MOF-74 were very dependent on the loading amount of AB. Other MOFs such as Ni-MOF-74 and HKUST-1 were also tried in this work. However, the structures collapsed with AB loading, which enabled them to be unsuitable for serving as a nanoscaffold. In our related study [25], new tactics for loading AB in MIL-101 (AB/MIL-101) or Ni-modified MIL-101 (AB/Ni@MIL-101) were proposed through wet impregnation. As shown in Figure 4, the dehydrogenation of AB/MIL-101 proceeded without an induction period or any detectable borazine and diborane, but with some ammonia. For the case of AB/Ni@MIL-101, the dehydrogenation peak for the most rapid rate was reduced to a lower temperature of 75 °C. Furthermore, the detrimental by-products of borazine, diborane, and ammonia were not detected during the whole experiments. Theoretical calculations indicated the establishment of B–O and Cr–N bonds and the existence of strong interaction between MIL-101 and NH_2BH_2 species, which could facilitate the dehydrogenation.

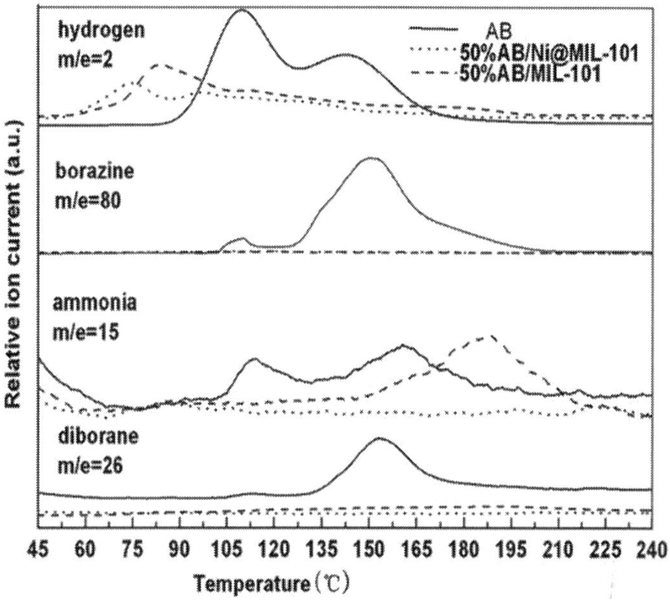

Figure 4. TPD-MS of AB, AB/Ni@MIL-101, and AB/MIL-101. Reproduced from Ref. [25] with permission.

3. Nanocatalysis of AB

$Ni_{1-x}Pt_x$ (x = 0-0.12) hollow spheres were prepared with an assistance of a template called as poly(styrene-co-methacrylic acid) and then used as catalysts for generating hydrogen from AB [26]. The morphology and the composition of the obtained $Ni_{0.88}Pt_{0.12}$ sample were shown in Figure 5. There are both broken and intact hollow spheres appeared in the sample. TEM images revealed that the porous shells of hollow spheres with a thickness of 20-40 nm contained many smaller nanowhiskers and nanoparticles, which enabled the increase of specific surface area of hollow spheres. Then the hollow spheres with a chemical composition of $Ni_{0.88}Pt_{0.12}$ were introduced into a methanol solution of AB. The mixture was vacuumed to obtain a dry solid sample with homogeneous loading of about 1.8 wt% Pt. TPD tests in Figure 6 showed that the onset temperature for hydrogen release was reduced to about 56 °C and the peak shifted to about 100 °C.

A "co-precipitation" process was developed using a typical wet-chemical method for fabricating nanoalloy to catalyze AB dehydrogenation [27]. Typically, a mixture of 2.0 mol % of $CoCl_2$ and 10 mL of THF was stirred by ultrasound at room temperature. Then AB was introduced into the resultant THF solution, which was immediately distilled under a

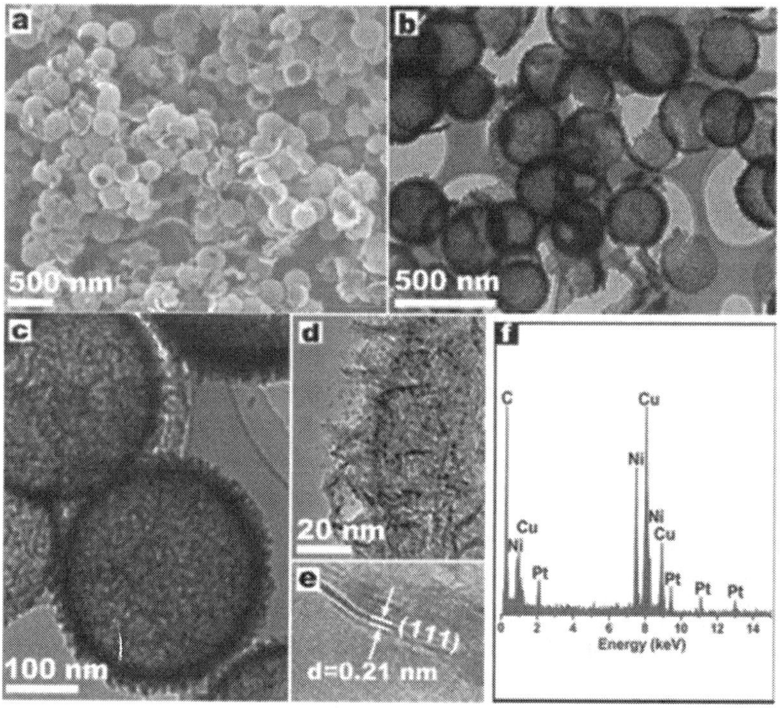

Figure 5. (a) SEM image, (b, c, d) TEM images, and (e) HRTEM image and (f) EDX spectrum of $Ni_{0.88}Pt_{0.12}$ hollow spheres. Reproduced from Ref. [26] with permission.

reduced pressure at room temperature. Finally, the Co-doped sample was obtained after continuous evacuation to eliminate the rest of THF. Ni-doped sample could be prepared in the same way. The size of most Ni and Co particles was below 3 nm, which indicated the peculiarity of the "co-precipitation" method in synthesizing nanoscaled metal catalysts. TPD-MS curves in Figure 7A showed that the catalyst-doped AB started to release H_2 at a very low temperature of 50 °C and exhibited broad dehydrogenation processes peaked at about 113 °C. Isothermal dehydrogenation test indicated that nearly 1 Eq of H_2 was released from the catalyst-doped AB at 59 °C without any induction period (Figure 7B). However, after being held at this temperature for more than 24 h, there was no evolution of any H_2 from pristine AB. Moreover, the detrimental by-product of borazine and sample forming were not observed during the catalytic dehydrogenation. Determined by the Kissinger model, the activation energy E_a was reduced to 117 and 123 kJ mol $^{-1}$ for Co- and Ni-doped AB. For the case of Co-doped AB, from electron paramagnetic resonance (EPR) characterization, Co^{2+} was partially reduced in the preparation, which was regarded as the catalytically active species for AB dehydrogenation.

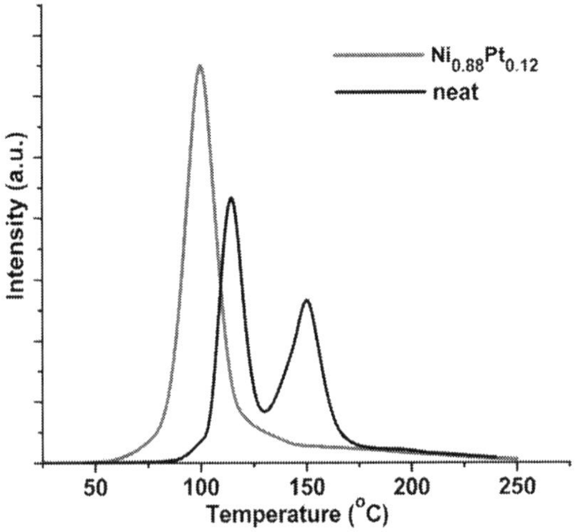

Figure 6. TPD-MS of AB mixed with $Ni_{0.88}Pt_{0.12}$ hollow spheres (red line). For comparison, neat AB (black line) was also included. Reproduced from Ref. [26] with permission.

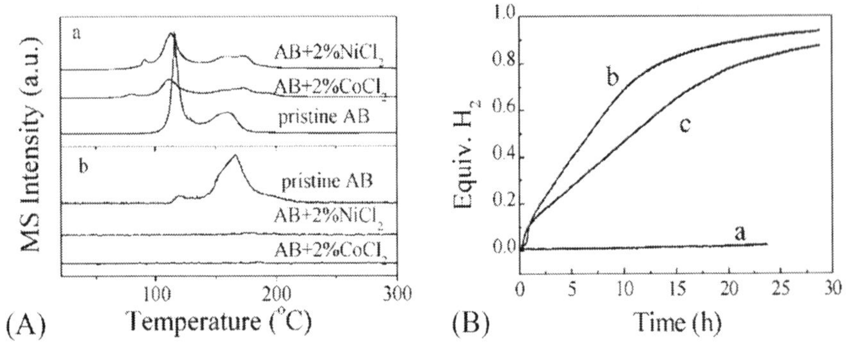

Figure 7. (A) TPD-MS of pristine and 2.0 mol% Co- or Ni-doped AB samples: (a) H_2 signal and (b) borazine signal. (B) Isothermal hydrogen release tests at 59 °C on (a) pristine, (b) Co-doped, and (c) Ni-doped AB samples. Reproduced from Ref. [27] with permission.

Because the chlorides of cobalt and nickel as precursors have better performance in catalyzing the dehydrogenation of AB, $FeCl_3$ and $FeCl_2$ were tried in the following studies [28,29]. Different amounts of $FeCl_3$ or $FeCl_2$ were introduced in solid AB using a "co-precipitation" method. Generally, $FeCl_3$ exhibited much better performance in catalyzing AB dehydrogenation than $FeCl_2$. Because of the formation of FeB nanoalloy in particle size of 2-5 nm, AB doped with 5 mol% $FeCl_3$ gave excellent performance with an onset temperature for hydrogen release at about 55 °C. Volumetric release tests (Figure 8) showed that $FeCl_3$-doped AB evolved H_2 immediately at 60 °C without an induction period observed. With the increase of temperature to 80 °C and 100 °C, it could release more than 1.2 and 1.5 Eq of H_2, respectively. Also, sample foaming upon heating and the by-products (borazine and ammonia) were significantly suppressed. More importantly, crystalline linear polyaminoborane (PAB) was found upon dehydrogenation of $FeCl_3$-doped AB, which was the first case on the preparation of crystalline PAB through solid-state reaction. Theoretical calculations showed that the growth of crystalline PAB may follow the dehydrogenation-chain growth mechanism. When the loading amount of $FeCl_3$ was reduced to 2.0 mol%, the optimal performance of the formation of crystalline PAB was achieved.

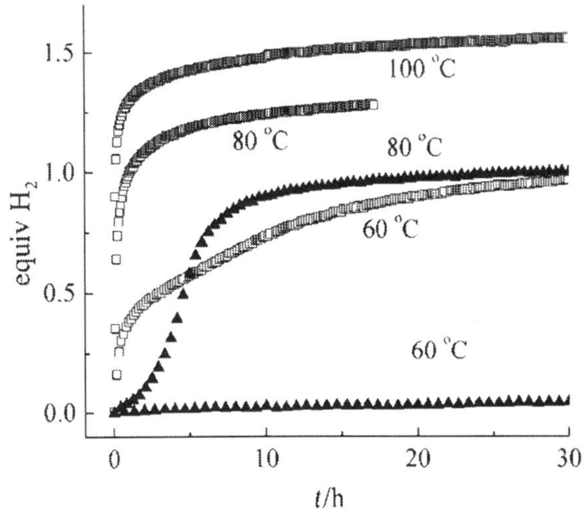

Figure 8. Volumetric release of 5.0 mol% $FeCl_3$-doped AB samples (□). For comparison, pristine AB (▲) was also included as a reference. Reproduced from Ref. [28] with permission.

As shown in Figure 9, a nanocomposite of AB/Li-CMK-3 was successfully synthesized by an impregnation method, i.e., AB loaded into a Li-doped CMK-3 (a class of carbon material with an ordered mesoporous structure) [30]. It could release more than 7 wt% H_2 at 60 °C, which was ascribed to the synergy between catalytic effect and nanoconfinement in AB/Li-CMK-3 system. These effects resulted in a significant enhancement of the dehydrogenation kinetics and a great suppression of detrimental by-products. In reference [31], two novel catalysts based on MOFs (denoted as MOF1cat and MOF2cat) were developed, in which in situ-formed metal Ni and the retained frameworks greatly promoted their catalytic activity. Loading with only 1.0 mol% catalyst in AB would give rise to higher dehydrogenation capacity, more rapid hydrogen releasing rates, lower onset temperature, and much less reaction exothermicity. At 90 °C, about 8 wt% H_2 could be released from catalyzed samples. The reaction enthalpies of

AB dehydrogenation catalyzed by MOF1cat and MOF2cat were determined to be −4.3 and −7.9 kJ mol^{-1}, respectively, much less than −21 kJ mol^{-1} for pristine AB. The apparent activation energies of 131 and 160 kJ mol^{-1} were also much lower than 184 kJ mol^{-1} for pristine AB.

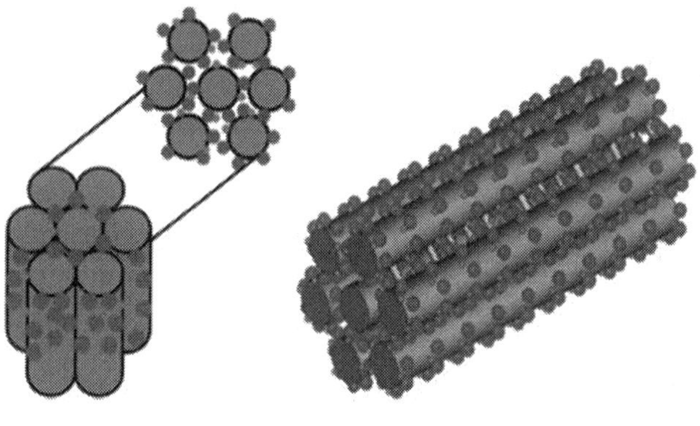

Figure 9. Schematic model of AB/Li-CMK-3 nanocomposites (AB: dark gray spheres; Li: light gray spheres; CMK-3 substrate: gray rod). Reproduced from Ref. [30] with permission.

Other catalysts consisting of noble metals were fabricated. For instance, the catalytic dehydrogenation properties of AB by carbon nanotubes functionalized with Pt nanoparticles (denoted as Pt@CNTs) were reported, in which Pt@CNTs was synthesized through an "ammonia-deliquescence" method [32]. From Figure 10, AB/Pt@CNTs had two peaks for hydrogen release at temperatures of about 108 °C and 150 °C. Furthermore, great suppression of borazine and no sample foaming and expansion were observed during the whole dehydrogenation. Of particular concern is achieving much less exothermicity for dehydrogenation. Meanwhile, AB/Pt@CNT had ameliorative dehydrogenation kinetics, i.e., releasing 1 Eq of H_2 in 5 h at 70 °C, which was evidenced by the reduced activation energy of 106.2 kJ mol^{-1}. It was also indicated that the combination of the nanoconfinement of AB into CNTs and the synergetically catalytic effects of Pt nanoparticles and CNTs were primarily responsible for enhancing the dehydrogenation properties. In addition, Pd, Pt, and Ni nanoparticles well deposited on MCM-48 (a silica-based material with mesoporous structure) were developed by a magnetron sputtering technique [33]. Then the resultant catalysts were introduced into AB by impregnation method using anhydrous diethyl ether as solvent and displayed obvious catalytic effects for hydrogen release from AB. For example, the onset temperature for dehydrogenation was about 93 °C for AB/MCM-48-Pd (weight ratio of 2:1), with a shift of 12 °C to reduced temperature compared with pristine AB. For all samples catalyzed by well-dispersed Pd, Pt, and Ni nanoparticles on MCM-48, borazine as a detrimental by-product was significantly suppressed, and there was no foaming and expansion during dehydrogenation process.

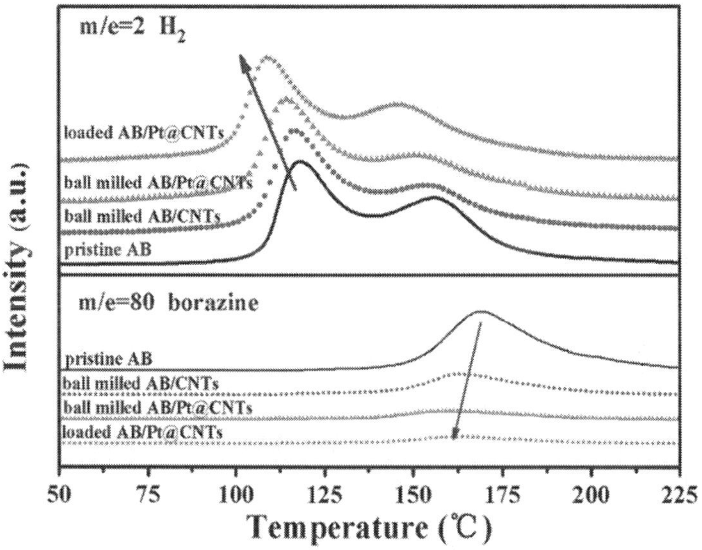

Figure 10. TPD-MS of pristine AB and loaded AB/Pt@CNTs. Reproduced from Ref. [32] with permission.

4. Metal amidoboranes, MABs

In 1938, sodium amidoborane ($NaNH_2BH_3$, NaAB, similarly hereinafter for other amidoboranes) was initially prepared by a reaction of borane etherate with sodium and ammonia [34]. For the case of lithium amidoborane ($LiNH_2BH_3$, LiAB), it was synthesized through deprotonation of AB in THF using *n*-butyllithium at 0 °C and then used as a reducing reagent [35]. However, the crystal structure of LiAB and NaAB was not reported until 2008 [36-38]. A solid-state synthetic method was developed for MABs through controlled mechanical ball milling AB with MH (M = Li or Na) in an equivalent molar amount. Calcium amidoborane ($Ca(NH_2BH_3)_2$, CaAB) was prepared by a wet-chemical reaction between AB and CaH_2 in THF [39]. As a matter of fact, CaAB·2THF rather than pure CaAB was obtained in final product due to the strong coordination of THF with Ca^{2+}. Wu et al. [38] synthesized solvent-free CaAB by ball milling solid-state CaH_2 and AB. More recently, some new MABs such as potassium amidoborane (KNH_2BH_3, KAB) [40], strontium amidoborane ($Sr(NH_2BH_3)_2$, SrAB) [41], and yttrium amidoborane ($Y(NH_2BH_3)_3$, YAB) [42] were successfully synthesized. In general, there are two synthetic approaches for the preparation according to the reported MABs, i.e., solid-state mechanical ball milling and wet-chemical synthesis. In solid-state ball milling method, the chemical interaction between protic H and hydridic H was regarded as one of main driving forces for hydrogen release [43-45]. It should be pointed out that the milling conditions are crucial during the synthesis of MABs because of their lower temperatures for dehydrogenation. In a wet-chemical reaction, the adducts rather than pure MABs are always obtained because the solvent is easily coordinated with amidoborane, especially alkaline-earth metal amidoboranes [39]. Because of relatively strong combination between solvent molecules and metal cations, special procedures are needed for removal of solvent, during which the decomposition of MABs should be avoided.

4.1. Lithium amidoborane ($LiNH_2BH_3$, LiAB) and sodium amidoborane ($NaNH_2BH_3$, NaAB)

As mentioned above, LiAB or NaAB could be prepared by carefully milling a mixture of AB and an equivalent LiH or NaH. LiAB crystallizes in an orthorhombic cell (space group *Pbca*) with lattice parameters of a = 7.11274(6) Å, b = 13.94877(14) Å, and c = 5.15018(6) Å (Figure 11) [36]. Moreover, NaAB is an isostructural compound with LiAB and the lattice parameters are a = 7.46931(7) Å, b = 14.65483(16) Å, and c = 5.65280(8) Å [36]. In their crystal structures, the bond lengths of Na-N and Li-N are 2.35 and 1.98 Å. The substitution of one H atom in NH_3 group by Li^+ gave rise to the slight shortening of bond length for B-N (1.56 Å) compared to that in AB (1.58 Å) [46]. Moreover, the B-H distance is 1.079 Å in LiAB and 1.245 Å averagely in NaAB, which are different from that that in AB (1.11 Å) [47]. The establishment of an ionic Li-N or Na-N bond and the resulting reinforce of B-N bond in $[NH_2BH_3]^-$ group in LiAB or NaAB resulted in an improved dehydrogenation properties. LiAB and NaAB started to release H_2 at 85 °C and had a peak at about 90 °C [36,37]. In addition, no borazine was detectable during the dehydrogenation. Furthermore, the heat for dehydrogenation from LiAB and NaAB was determined to be ~-3 and -5 kJ (mol H_2^{-1}), remarkably less exothermic than that of pristine AB (-21 kJ (mol H_2^{-1})) [4]. At 91 °C, about 7.4 and 11 wt% H_2 could be released from NaAB and LiAB, respectively (Figure 12).

LiAB could also be synthesized by reaction of LiH and AB in THF after the evolution of 1 Eq of H_2. Surprisingly, it then releases another 2 Eq of H_2 in THF at a lowest temperature of 40 °C reported so far [48]. The dehydrogenation of LiAB in THF was determined to be zero-order based on the concentration of LiAB. However, the dehydrogenation mechanism is still unclear and needs further investigations. Through an electrospinning technique, LiAB nanoparticles were well dispersed in carbon nanofibers (denoted as LiAB@CNFs), which had onset and peak temperatures of about 40 °C and 80 °C, significantly lower than pristine LiAB. Upon heating to 100 °C, all hydrogen in LiAB of 10.6 wt% could be released in only 15 min [49]. For the case of NaAB, it could also be prepared through a wet-chemical method by reacting AB with NaH at -3 °C or with $NaNH_2$ at 25 °C in THF [50]. After dehydrogenation of LiAB or NaAB, the solid residues were found to be poorly crystalline phases. A following study indicated that solid NaH was detected for NaAB [50]. However, a recent investigation on the post-dehydrogenated products of NaAB at 200 °C suggested that except NaH, an amorphous phase with the chemical composition of $Na_{0.5}NBH_{0.5}$ (or regarded as a mixture of NaNBH and *h*-BN) was identified by NMR technique [51]. Unfortunately, the results from Fijałkowski et al. [52] showed that a substantial amount of NH_3 was released during synthesis by ball milling and the following thermal decomposition of NaAB.

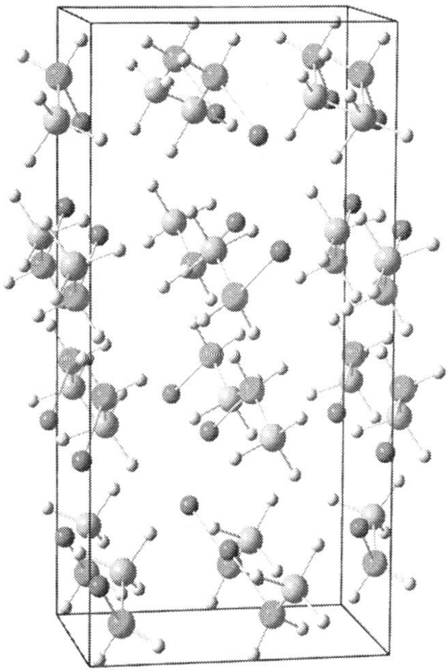

Figure 11. Crystal structure of LiAB. B: orange spheres; N: green spheres; H: white spheres; and Li: red spheres. Reproduced from Ref. [36] with permission.

4.2. Lithium amidoborane-ammonia borane ($LiNH_2BH_3 \cdot NH_3BH_3$, LiAB·AB) and β-LiAB

Careful exploration by XRD technique during ball milling of AB and LiH (in 1:1 molar ratio) indicated a stepwise process in LiAB formation [53]. A new complex hydride of lithium amidoborane-ammonia borane ($LiNH_2BH_3 \cdot NH_3BH_3$, LiAB·AB) was formed as an intermediate [38,53]. Pure LiAB·AB could be prepared by ball milling a mixture of AB/LiAB or 2AB/LiH [54]. It crystallizes in a monoclinic cell (space group *P21/c*) with lattice parameters of a = 7.0536(9) Å, b = 14.8127(20) Å, c = 5.1315(7) Å, and β = 97.491(5)°. There is the intergrowth of AB and LiAB layers in crystal structure of LiAB·AB. As shown in Figure 13, each Li^+ is bonded with one N atom from $[NH_2BH_3]^-$ with a Li-N bond in 2.05 Å and coordinated with hydrogen atoms from BH_3 group in two LiAB and one AB. The dihydrogen bonding of NH···HB is 1.902 Å in AB layer. About 14.0 wt% H_2 could be released at 228 °C from LiAB·AB after its melting with an onset temperature of 58 °C. Furthermore, another allotrope of β-LiAB was found during ball milling process. Interestingly, it transformed from initially formed α-LiAB and could transform to α-LiAB upon extended milling [53]. β-LiAB also crystallizes in an orthorhombic cell (space group *Pbca*) with lattice parameters of a = 15.15 Å, b = 7.726 Å, and c = 9.274 Å. However, its unit cell has a double volume of α-LiAB, in which two alternative LiAB layers were observed along the a axis. The coordination of Li^+ was found to be distortedly tetrahedral

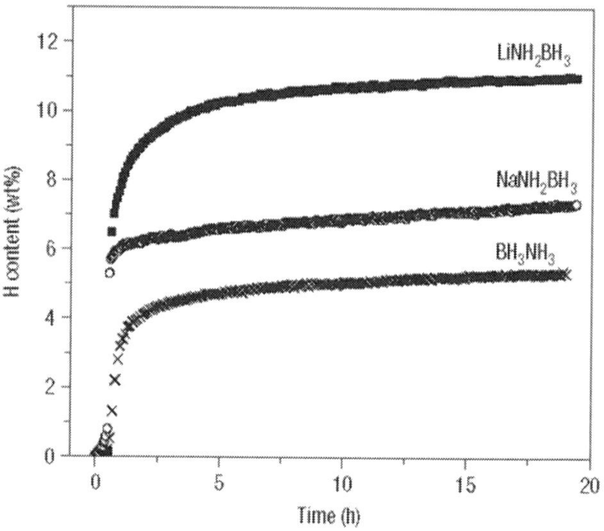

Figure 12. Isothermal dehydrogenation of LiAB, NaAB, and post-milled AB samples at about 91 °C. Reproduced from Ref. [36] with permission.

in β-LiAB due to different distances for Li1-B and Li2-B (Figure 13b). The bond length of Li-N (1.93 and 2.04 Å for Li1-N and Li2-N) is similar to that in α-LiAB (1.98 [36] or 2.02 Å [53]). Both allotropes of LiAB displayed identical dehydrogenation behaviors, i.e., 10.8 wt% H_2 could be released at 180 °C [53].

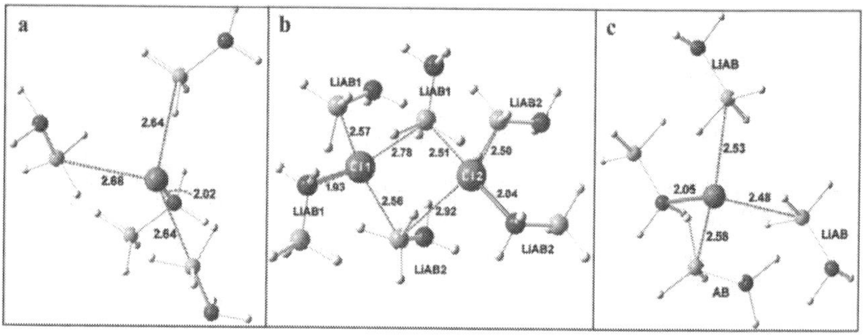

Figure 13. Coordination of Li⁺ in (a) α-LiAB, (b) β-LiAB, and (c) LiAB·AB. Li: purple spheres; B: brown spheres; N: blue spheres; and H: white spheres. Reproduced from Ref. [53] with permission.

http://dx.doi.org/10.5772/61048

4.3. Potassium amidoborane (KNH_2BH_3, KAB)

Through the reaction of AB with an equivalent KH, polycrystalline KAB was synthesized in THF, while single crystal was obtained in a mixed solvent of diglyme and hexane at room temperature [40]. Similar to LiAB and NaAB, KAB crystallizes in an orthorhombic cell (space group *Pbca*) with lattice parameters of a = 9.4304(1) Å, b = 8.26112(1) Å, and c = 17.3403(2) Å. However, different from Li^+ in LiAB and Na^+ in NaAB, K^+ is octahedrally coordinated by $[NH_2BH_3]^-$ groups with the formation of three K-N bonds (3.0207-3.1345 Å) and three $K\cdots BH_3$ coordinations (Figure 14A). The distance of closest $NH\cdots HB$ (2.265 Å) in KAB is smaller than those in LiAB (2.372 Å) and NaAB (2.717 Å). As shown in Figure 14B, an endothermic peak for melting was observed before an exothermic event for hydrogen release. It could release 1.5 Eq of H_2 (6.5 wt%) at ~80 °C. With the temperature increase to 160 °C, another 0.5 Eq of H_2 was released. No ammonia or borazine was detected throughout the dehydrogenation.

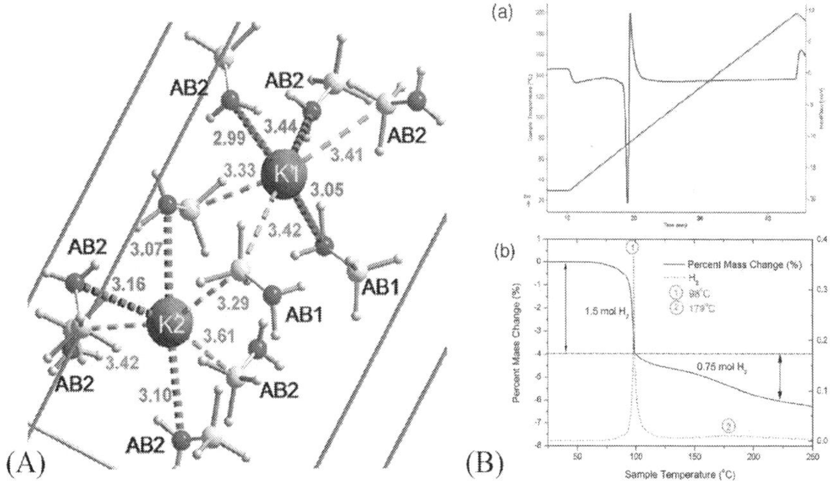

Figure 14. (A) Crystal structure of KAB. K: green spheres; N: blue spheres; B: cream spheres; and H: white spheres. (B) (a) DSC and (b) IGA data for KAB decomposition. Reproduced from Ref. [40] with permission.

4.4. Magnesium amidoborane ($Mg(NH_2BH_3)_2$, MgAB)

MgAB is supposed to be a promising material for storing hydrogen because it has a hydrogen content of 11.8 wt%. However, it is unsuccessful to prepare crystalline MgAB either by ball milling method or *via* wet-chemical method until the beginning of 2013. MgAB was synthesized through aging treatment of post-milled $2AB/MgH_2$ or 2AB/Mg powder, in which the formation of MgAB was not a straightforward process but experienced a prerequisite step of phase transition [55]. However, the crystal structure was failed to solve. In 2010, its structure was successfully proposed through theoretical simulation [15]. MgAB has a monoclinic cell (space group *C*2) with lattice parameters of a = 8.5772 Å, b = 5.6048 Å, c = 5.6216 Å, and β = 85.8476°. Two Mg-N bonds with the length of 2.111 Å and two Mg···BH_3 coordinations are around Mg^{2+} (Figure 15). As a stable compound at room temperature, MgAB could release ~10 wt% H_2 at 300 °C, without volatile by-products detectable [55].

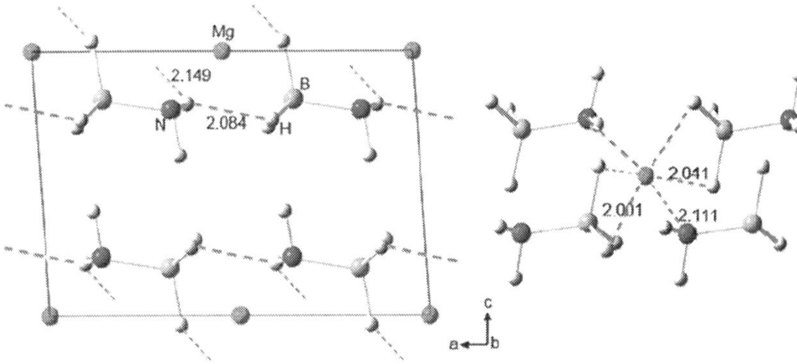

Figure 15. Simulated crystal structure of MgAB. Mg: green spheres; N: blue spheres; B: orange spheres; and H: white spheres. Reproduced from Ref. [15] with permission.

4.5. Calcium amidoborane ($Ca(NH_2BH_3)_2$, CaAB)

Due to the strong coordination between solvent and metal cation, CaAB·2THF adduct rather than pure CaAB was first synthesized through reacting CaH_2 with two equivalent AB in THF [39]. The specially octahedral coordination of Ca^{2+} with two NH_2, two BH_3, and two O resulted in an extended chain-like structure. After desiccation under a dynamic flow of argon, about 10% remaining THF was evolved from 70 °C to 105 °C, and hydrogen was released mainly from 120 to 245 °C. Another adduct of [(DIPP-nacnac)$CaNH_2BH_3(THF)_2$] was also reported [56], in which Ca^{2+} is octahedrally coordinated to DIPP-nacnac [(2,6-$iPr_2C_6H_3$)NC(Me)C(H)C(Me)N(2,6-$iPr_2C_6H_3$)] bidentate ligand, $[NH_2BH_3]^-$, THF. Using mechanical milling method from CaH_2 and two equivalent AB, Wu et al. [38] synthesized a solvent-free CaAB, which crystallizes in a monoclinic cell (space group *C*2) with lattice parameters of a = 9.100(2) Å, b = 4.371(1) Å, c = 6.441(2) Å, and β = 93.19°. Ca^{2+} is octahedrally coordinated through two Ca-N bonds with the length of ~2.466 Å and four coordinations of Ca···BH_3 with the distance in the range of 2.87-3.03 Å (Figure 16). Different from the CaAB adducts, there are two steps for hydrogen release with peaks at 100 and 140 °C. Upon heating to 250 °C, 2 Eq of H_2 was released with the solid residues in amorphous state.

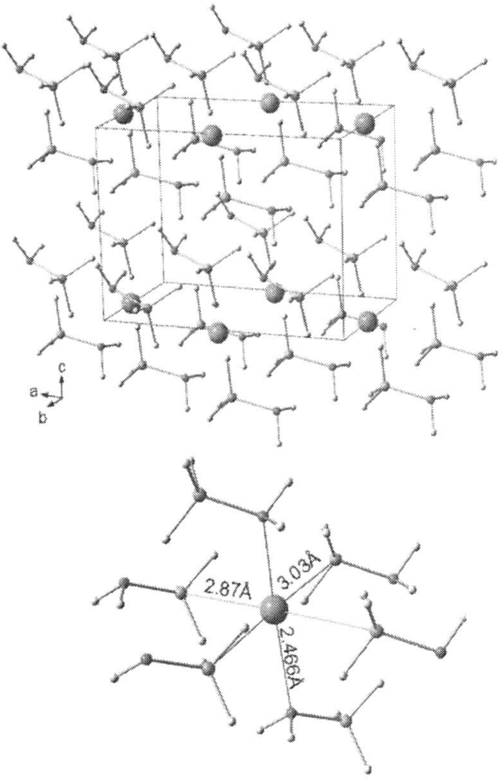

Figure 16. (Top) Crystal structure of CaAB. Ca: orange spheres; B: green spheres; N: blue spheres; and H: white spheres. (Bottom) Coordination environment of Ca^{2+}. Reproduced from Ref. [38] with permission.

4.6. Strontium amidoborane ($Sr(NH_2BH_3)_2$, SrAB)

Through reacting SrH_2 with AB in a 1:2 molar ratio, SrAB was successfully synthesized by moderate milling and the following heating treatment at 45 °C [41]. Similar to CaAB, SrAB also crystallizes in a monoclinic cell (space group C2) with lattice parameters of a = 8.1660(4) Å, b = 5.0969(3) Å, c = 6.7258(4) Å, and β = 94.392(4)°. Sr^{2+} is octahedrally coordinated by $[NH_2BH_3]^-$ groups to form two Sr-N bonds with a length of 2.68 Å and four $Sr\cdots BH_3$ coordinations (Figure 17). Upon heating, SrAB started to decompose into H_2 and $Sr(NBH)_2$ at about 60 °C. In an isothermal dehydrogenation performed at a temperature of 80 °C, 4 Eq of H_2 could be rapidly released in a few minutes. However, the evolved hydrogen was contaminated by undesirable by-products of NH_3 and B_2H_6 evolved from the decomposition of the resulting $Sr(NBH)_2$.

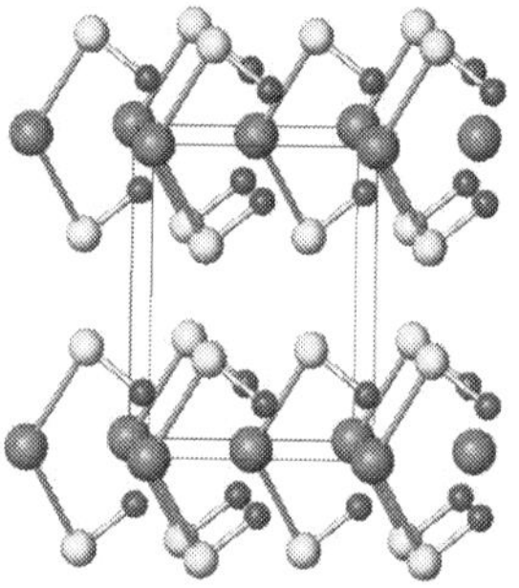

Figure 17. Crystal structure of SrAB. Sr: red spheres; N: yellow spheres; and B: blue spheres. Reproduced from Ref. [41] with permission.

4.7. Yttrium amidoborane ($Y(NH_2BH_3)_3$, YAB)

Ball milling YCl_3 with three equivalent LiAB gave rise to the formation of YAB [42]. In fact, it was a homogenous mixture with a side product of LiCl, which is very difficult to separate. YAB has a monoclinic cell (space group *C*2/*c*) with lattice parameters of a = 13.18902(63) Å, b = 7.82233(38) Å, c = 14.874274(68) Å, and β = 92.42620(40)°. However, it was thermodynamically unstable and spontaneously decomposed in several days at room temperature. A large amount of H_2 with NH_3 impurity was released from freshly prepared sample upon heating to 250 °C. Motivated by the synthesis of YAB, the attempt to synthesize iron (III) amidoborane ($Fe(NH_2BH_3)_3$, FeAB) through metathesis of $FeCl_3$ and three equivalent LiAB in THF was unsuccessful [57]. Instead of FeAB formation, 1.5 Eq of H_2 was released and a black solid containing LiCl was produced. Meanwhile, Fe^{3+} was reduced during dehydrogenation evidenced by Mössbauer and XAFS characterizations.

4.8. $Na[Li(NH_2BH_3)_2]$

$Na[Li(NH_2BH_3)_2]$ was reported to be the first mixed-cation amidoborane, which was prepared by milling a mixture of NaH/LiH/2AB performed on a high-energy disk mill [58]. It crystallizes in a triclinic cell (space group $P\bar{1}$) with lattice parameters of a = 5.0197(4) Å, b = 7.1203(7) Å, c = 8.9198(9) Å, α = 103.003(6)°, β = 102.200(5)°, and γ = 103.575(5)° [58]. As shown in Figure 18B, Na^+ is octahedrally coordinated with H atoms from BH_3 groups with different Na-H distances, while Li^+ is tetrahedrally coordinated by one H atom from BH_3 group and three N atoms from NH_2 groups. In addition, the formed chains of dimeric $[Li(NH_2BH_3)_2]_2^-$ anions linked through Li···H coordinations were found in $Na[Li(NH_2BH_3)_2]$. Thus, $Na[Li(NH_2BH_3)_2]$ was regarded as an ionic compound composed of Na^+ and $[Li(NH_2BH_3)_2]_2^-$, which is totally different from NaAB and LiAB. $Na[Li(NH_2BH_3)_2]$ decomposed exothermally from 75 °C to 110 °C with 6.0 wt% H_2 released. An additional 3.0 wt% H_2 could be released in the temperature range of 110-200 °C. During its decomposition, there was an evolution of ammonia [58]. One year later, Li et al. [59] prepared $Na[Li(NH_2BH_3)_2]$ through a wet-chemical method, i.e., recrystallizing a mixture of LiAB/NaAB (1:1 molar ratio) or reacting AB with a post-milled mixture of LiH/NaH (1:1 molar ratio) in THF. However, $Na[Li(NH_2BH_3)_2]$ was theoretically determined to be in an orthorhombic cell (space group of $Pca2_1$) with lattice parameters of a = 14.639 Å, b = 7.169 Å, and c = 5.367 Å, which is significantly different from the structure determined experimentally. For example, the coordination environments of Li^+ and Na^+ are similar in its simulated structure, i.e., surrounded by one N atom from $[NH_2BH_3]^-$ and six H atoms from $[NH_2BH_3]^-$ (Figure 18A). Furthermore, the calculated free energy $\Delta G(x)$ indicated that all of the studied compositions in $Na_xLi_{1-x}NH_2BH_3$ except x = 0.125 were also thermodynamically favored [59].

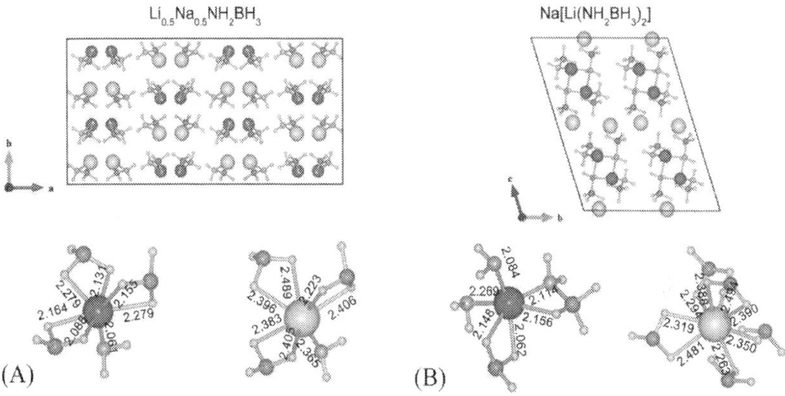

Figure 18. Crystal structure of $Na[Li(NH_2BH_3)_2]$ determined theoretically (A) and experimentally (B). Coordinations of Li^+ and Na^+ are displayed under each structure. Li: big red spheres; Na: big yellow spheres; N: small light blue spheres; B: small green spheres; and H: small pink spheres. Reproduced from Ref. [59] with permission.

4.9. $NaMg(NH_2BH_3)_3$

Almost at the same time, another dual-metal (alkali metal and alkaline-earth metal) amidoborane, namely, $NaMg(NH_2BH_3)_3$, was reported [60]. It was prepared by mechanically milling $3AB/NaMgH_3$ powdery mixture and the following treatment at 45 °C according to the reaction equation: $3NH_3BH_3 + NaMgH_3 \rightarrow NaMg(NH_2BH_3)_3 + 3H_2$. From its SR-XRD pattern, $NaMg(NH_2BH_3)_3$ was ascribed to a monoclinic cell (space group $P2_1$ or $P2_1/m$) with lattice parameters of a = 17.011 Å, b = 9.432 Å, c = 9.398 Å, and β = 115.99°. However, the detailed crystal structure was unsuccessfully solved in reference [60]. For the post-milled $3AB/NaMgH_3$ without heating treatment, there were two steps for dehydrogenation. It started to release H_2 exothermically at around 50 °C for the first step with yielding about 3.5 wt% H_2, which agrees well with the amount of hydrogen released during heating treatment at 45 °C. Therefore, it could be concluded that the first step for dehydrogenation gave rise to the formation of $NaMg(NH_2BH_3)_3$ and the second one was for its own dehydrogenation with the beginning at about 140 °C and the most rapid release rate at about 170 °C. No volatile by-products and sample foaming were observed during the decomposition process upon heating to 250 °C. It is worth noting that the dehydrogenation was determined to be of mild endo-thermicity of 3.4 kJ (mol $H_2)^{-1}$, which is the first case reported so far and entirely different from the exothermic dehydrogenation for monometal amidoboranes of NaAB and MgAB. This indicates that dual-cation combination in amidoboranes plays a vital role in altering their dehydrogenation thermodynamics. From thermodynamic point of view, the endothermic dehydrogenation provides a possibility to achieve rehydrogenation at moderate conditions. However, in this study, it was unsuccessful to recharge the solid residual at moderate conditions.

4.10. $Na_2Mg(NH_2BH_3)_4$

$Na_2Mg(NH_2BH_3)_4$ was first synthesized by ball milling a mixture of $NaH/MgH_2/2AB$ [61]. It was identified to have a space group of $I4_1/a$ and approximate lattice parameters of $a = 9.415$ Å and $c = 12.413$ Å. The crystal structure illustrated in Figure 19 shows that Mg^{2+} is coordinated with four $[NH_2BH_3]^-$ groups *via* Mg-N bonds (2.104 Å) to form $[Mg(NH_2BH_3)_4]^{2-}$ tetrahedron. Na^+ is octahedrally coordinated only with six BH_3 groups with the distances of Na-B ranging from 2.900 to 3.634 Å. The distances of Na-H coordinations between Na^+ and its close hydridic H in BH_3 groups are in the range of 2.383-2.943 Å. As a consequence, Mg^{2+} and Na^+ are linked through different coordinations with both ends of the bridging $[NH_2BH_3]^-$ groups and form an ordered structure in coordinations. $Na_2Mg(NH_2BH_3)_4$ had an onset temperature for dehydrogenation at ~65 °C and a total amount of 8.4 wt% H_2 was released upon heating to 200 °C, with contamination by trace amount of borazine and ammonia. In another paper [62], a wet-chemical method was proposed to synthesize pure $Na_2Mg(NH_2BH_3)_4$. Typically, a solid mixture of $Mg(NH_2)_2/2NaH/4AB$ was introduced into THF. After 0.7 Eq of gaseous products (H_2 and NH_3) was released, pure $Na_2Mg(NH_2BH_3)_4$ as the precipitate was obtained through filtration, which agrees well with the one synthesized by ball milling method [61].

4.11. $K_2Mg(NH_2BH_3)_4$

$K_2Mg(NH_2BH_3)_4$ was synthesized by ball milling method according to the reaction equation: $2Mg(NH_2BH_3)\cdot NH_3 + 2KH \rightarrow K_2Mg(NH_2BH_3)_4 + Mg(NH_2)_2 + 2H_2$ [62]. Similarly, $Na_2Mg(NH_2BH_3)_4$ could also be prepared through this procedure. However, as a by-product, $Mg(NH_2)_2$ is very difficult to separate or remove. $K_2Mg(NH_2BH_3)_4$ crystallizes in a tetragonal cell (space group $I4_1/a$) with lattice parameters of $a = b = 9.5974(17)$ Å and $c = 13.581(4)$ Å. As shown in Figure 20, Mg^{2+} is tetrahedrally coordinated with four N atoms (2.115 Å for Mg–N bond length) from four $[NH_2BH_3]^-$ groups to form $[Mg(NH_2BH_3)_4]^{2-}$ species. Unlike Mg^{2+}, without the formation of K–N bond, K^+ is octahedrally coordinated by six $[Mg(NH_2BH_3)_4]^{2-}$ groups through six $K\cdots BH_3$ coordinations with K-B distances ranging from 3.361 to 3.617 Å. Due to the difficulties in the removal of $Mg(NH_2)_2$ from the products, the dehydrogenation

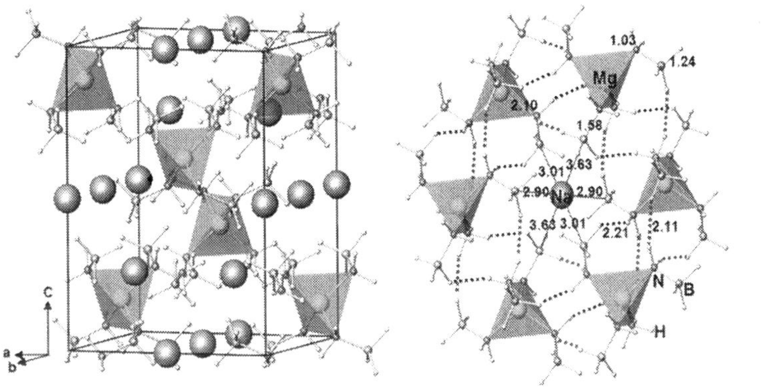

Figure 19. Crystal structure of $Na_2Mg(NH_2BH_3)_4$. Na: pink spheres; Mg: yellow spheres; N: blue spheres; B: orange spheres; and H: white spheres. Reproduced from Ref. [61] with permission.

was measured based on the composite of [$M_2Mg(NH_2BH_3)_4$ + $Mg(NH_2)_2$, (M = Na, K)] [62]. A very small amount of NH_3 was released from room temperature to 150 °C before dehydrogenation for both composites, with a mass loss of 3.7 and 3.8 wt%. Hydrogen was released with the most rapid rate at about 158 and 153 °C for [$K_2Mg(NH_2BH_3)_4$+$Mg(NH_2)_2$] and [$Na_2Mg(NH_2BH_3)_4$+$Mg(NH_2)_2$] composites, respectively. Most important of all, both composites exhibited the endothermic dehydrogenation in open system, i.e., 5.2 kJ (mol H_2)$^{-1}$ for [$Na_2Mg(NH_2BH_3)_4$+$Mg(NH_2)_2$] and 7.0 kJ (mol H_2)$^{-1}$ for [$K_2Mg(NH_2BH_3)_4$+$Mg(NH_2)_2$]. A total amount of about 7 and 7.8 wt% H_2 was achieved upon heating to 285 °C for these two composites. However, the role of $Mg(NH_2)_2$ remains unknown in inducing endothermic dehydrogenation of composites. Further research into the interaction between amide and dual-metal amidoborane is needed.

4.12. (Li,Al) amidoborane

Bimetallic (Li,Al) amidoborane was prepared by ball milling of $4AB/Li_3AlH_6$ mixture [63]. It was also observed after heating a post-milled mixture of $LiAlH_4$/AB at 170 °C [64]. The information of crystal structure was not reported in both references [63,64]. It could release about 10 wt% H_2 at 200 °C, with a significant suppression of volatile by-products. The post-dehydrogenated products of (Li,Al) amidoborane could be regenerated through reduction by hydrazine in liquid ammonia. Around 3.5 wt% H_2 was released upon heating to 400 °C, corresponding to a 35% regeneration yield [63].

5. Metal amidoborane ammoniates

Ammonia was a hydrogen-rich compound (17.7 wt%). As a strong Lewis base, it is used as a ligand to form amine complexes with various compounds for hydrogen storage [65-69]. It is

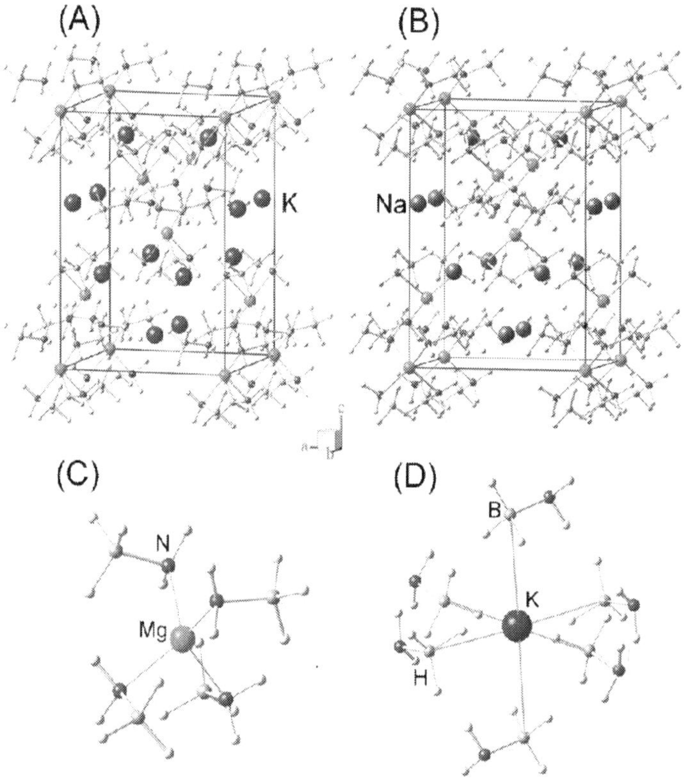

Figure 20. Crystal structure of (a) $K_2Mg(NH_2BH_3)_4$, (b) $Na_2Mg(NH_2BH_3)_4$, (c) Coordination of Mg^{2+}, and (d) Coordination of K^+. Reproduced from Ref. [62] with permission.

noted that because of the difference in electronegativity, positively charged H in NH_3 can interact with protic H in metal amidoboranes and borohydrides for facilitating hydrogen release.

5.1. Lithium amidoborane ammoniate ($LiNH_2BH_3 \cdot NH_3$, $LiAB \cdot NH_3$)

A new hybrid material with the formula of $LiNH_2BH_3NH_3$ was produced from the mixed AB and $LiNH_2$ in a 1:1 molar ratio, which can be treated as a monoammoniate of LiAB. It rapidly released 11.9 wt% H_2 under 250 °C, with the most rapid rate observed at 60 °C [70]. In fact, $LiAB \cdot NH_3$ was synthesized by reacting LiAB with an equivalent NH_3 [71]. At room temperature, it is a sticky matter in an amorphous state. Upon cooling to −20 °C, $LiAB \cdot NH_3$ becomes a solid and crystallizes in an orthorhombic cell (space group *Pbca*), which has the lattice parameters of $a = 9.711(4)$ Å, $b = 8.7027(5)$ Å, and $c = 7.1999(1)$ Å. However, the crystal structure was not reported in reference [71]. At room temperature, $LiAB \cdot NH_3$ decomposed into ammonia and LiAB under vacuum. However, at temperatures more than 40 °C, favorable dehydrogenation was observed. Volumetric release test showed that about 6.2 wt% H_2 could be released at 70 °C under argon, while 11.2 wt% H_2 at 60 °C under ammonia.

5.2. Calcium amidoborane diammoniate ($Ca(NH_2BH_3)_2 \cdot 2NH_3$, $CaAB \cdot 2NH_3$)

In 2009, Chua et al. [72] reported the synthesis, structure, and dehydrogenation of $CaAB \cdot 2NH_3$. They provided two synthesizing methods, i.e., ball milling of $Ca(NH_2)_2$/2AB mixture and reacting CaAB with two equivalent NH_3. It crystallizes in an orthorhombic cell (space group *Pna*21) with lattice parameters of $a = 18.673(3)$ Å, $b = 5.2283(8)$ Å, and $c = 8.5748(12)$ Å. Figure 21 shows that each Ca^{2+} has octahedral coordination through two $Ca \cdots NH_3$ coordinations (2.517 and 2.521 Å), two $Ca \cdots BH_3$ coordinations (2.650 and 2.807 Å), and two Ca-N bonds (2.397 and 2.521 Å), which results in a bonding network of $NH \cdots HB$ dihydrogen due to NH_3 coordinations. $CaAB \cdot 2NH_3$ decomposed into CaAB and 2 Eq of NH_3 under 100 °C in an open system. On the contrary, in a closed reactor, H_2 rather than NH_3 was released predominantly due to the existence of NH_3 equilibrium pressure, with an onset temperature at ~70 °C. Almost 9 wt% H_2 could be released upon heating to 300 °C. Compared to CaAB, the onset temperature has a 50 °C reduction and the dehydrogenation capacity at 300 °C has a 2 wt% increase.

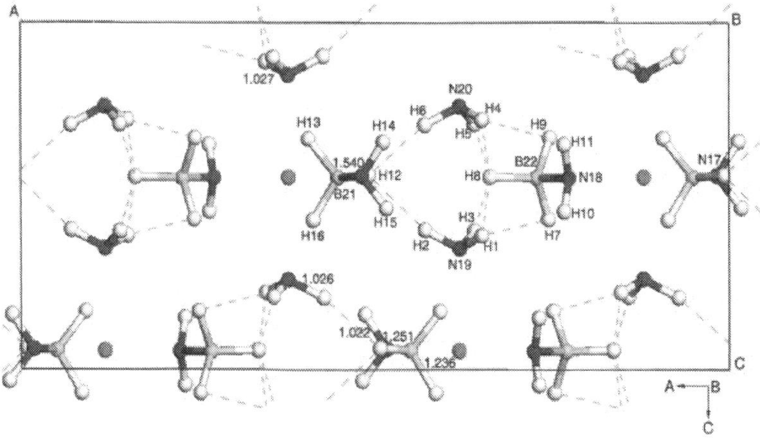

Figure 21. Crystal structure of $CaAB \cdot 2NH_3$. Ca: green spheres; N: blue spheres; B: pink spheres; and H: white spheres. Reproduced from Ref. [72] with permission.

5.3. Calcium amidoborane monoammoniate ($Ca(NH_2BH_3)_2 \cdot NH_3$, $CaAB \cdot NH_3$)

$CaAB \cdot NH_3$ was prepared by ball milling a mixture of CaNH/2AB [73]. Alternatively, ball milling a mixture of $Ca(NH_2)_2$/CaH_2/4AB could also give rise to the formation of $CaAB \cdot NH_3$

with accompanying 2 Eq of H_2 released. It crystallizes in a monoclinic cell (space group of $P2_1/c$) with lattice parameters of a = 10.5831(14) Å, b = 7.3689(11) Å, c = 10.2011(13) Å, and β = 120.796(6)° [73]. Unlike CaAB·2NH_3, Ca^{2+} in CaAB·NH_3 has a geometry of trigonal bipyramid (Figure 22), in which Ca^{2+} is surrounded by two NH_2 from $[NH_2BH_3]^-$ groups through Ca-N bonds (2.400 and 2.465 Å), two BH_3 from $[NH_2BH_3]^-$ groups through Ca···BH_3 interactions (2.669-3.182 Å), and one NH_3 through Ca···NH_3 interaction (2.472 Å). The special structure resulted in the network of NH···HB dihydrogen bond. TG-DSC-MS tests revealed that NH_3 was predominantly released under 100 °C with a weight loss of 12.6%, while 6.2 wt% H_2 was solely released from 100 to 300 °C. Compared with the theoretical value for deammoniation of CaAB·NH_3 (~14.5 wt%) and for dehydrogenation of CaAB (~8 wt%), some NH_3 should participate in dehydrogenation under 100 °C, which is evidenced by the concurrent release of H_2 from MS. However, in a closed reactor, hydrogen release in two steps was observed with an exothermic event detected in the first step. At 300 °C, 6 Eq of H_2 (~10.2 wt%) could be released, in which the amount of by-product NH_3 is less than 0.1 Eq.

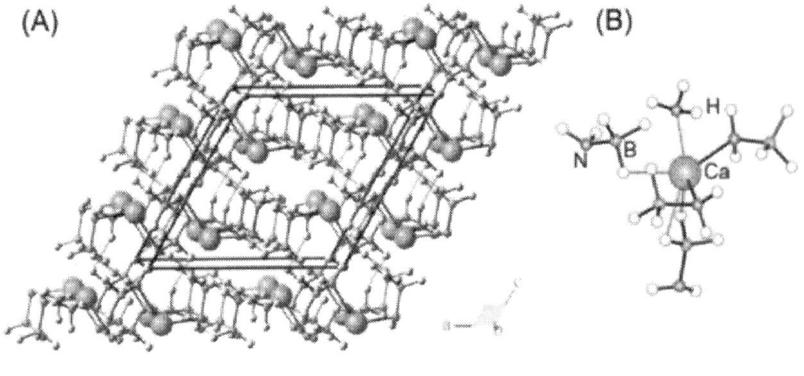

Figure 22. A) Crystal structure of CaAB·NH_3. (B) Coordination of Ca^{2+}. Ca: green spheres; N: blue spheres; B: pink spheres; and H: white spheres. Reproduced from Ref. [73] with permission.

5.4. Magnesium amidoborane monoammoniate ($Mg(NH_2BH_3)_2$·NH_3, MgAB·NH_3)

Ball milling a mixture of $Mg(NH_2)_2$/2AB could give rise to the formation of magnesium amidoborane diammoniate ($Mg(NH_2BH_3)_2$·2NH_3, MgAB·2NH_3). Unlike solid CaAB·2NH_3, MgAB·2NH_3 is a liquid matter at ambient condition. Upon releasing 1 Eq of NH_3, solid MgAB·NH_3 was formed [74]. Alternatively, MgAB·NH_3 could also be prepared by ball milling a mixture of MgNH/2AB [74]. In a monoclinic cell (space group of $P21/a$), MgAB·NH_3 has the lattice parameters of a = 8.8815(6) Å, b = 8.9466(6) Å, c = 8.0701(5) Å, and β = 94.0744(48)°. As shown in Figure 23, each Mg^{2+} has a tetrahedral coordination, in which Mg^{2+} forms two Mg-N bonds (2.104 and 2.129 Å) with the adjacent $[NH_2BH_3]^-$ groups, a Mg···NH_3 coordination (2.157 Å) with the adjacent NH_3, and a Mg···BH_3 coordination (2.126 Å) with the adjacent

$[NH_2BH_3]^-$ group. In $MgAB \cdot NH_3$, a dihydrogen bonding network was established through the interactions of NH···HB between NH_3 and BH_3 groups. Different from $CaAB \cdot 2NH_3$ and $CaAB \cdot NH_3$, the decomposition of $MgAB \cdot NH_3$ in an open system exhibited direct desorption of H_2 instead of NH_3, indicating easier interaction between $H^{\delta+}$ in NH_3 and $H^{\delta-}$ in BH_3 to the formation of H_2 or stronger Mg···NH_3 coordination. Hydrogen release of $MgAB \cdot NH_3$ started at about 50 °C, with 5.7 Eq of H_2 (11.4 wt%) released up to 300 °C. An amorphous solid residue with chemical composition of MgB_2N_3H was left. NH_3 was undetectable during the whole dehydrogenation process, which evidenced the complete conversion of NH_3.

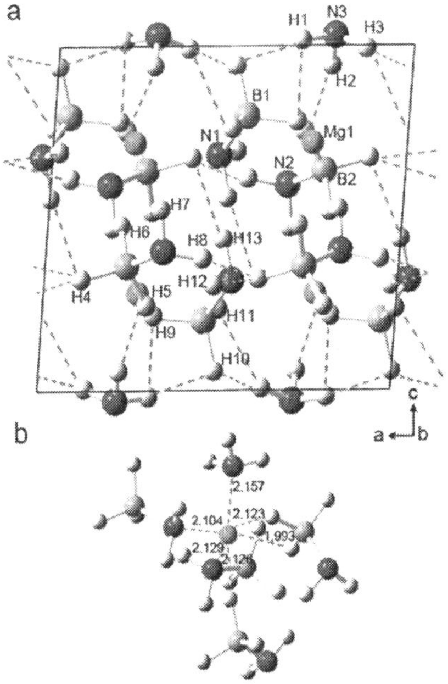

Figure 23. (a) Crystal structure of $MgAB \cdot NH_3$, and (b) Coordination of Mg^{2+}. Mg: green spheres; N: blue spheres; B: orange spheres; and H: white spheres. Reproduced from Ref. [74] with permission.

6. Metal borohydride-ammonia borane complexes

6.1. $2LiBH_4 \cdot NH_3BH_3$, $2LiBH_4 \cdot AB$

In 2010, the first metal borohydride-ammonia borane complex, namely, $2LiBH_4 \cdot AB$, was reported [75]. It was synthesized by ball milling AB with two equivalent $LiBH_4$ and had an orthorhombic structure (space group *Pnma*) with lattice parameters of a = 8.3118(8) Å, b = 12.428(1) Å, and c = 6.5944(7) Å. Figure 24 shows that there are alternating layers of borohydride and AB in $2LiBH_4 \cdot AB$. And Li^+ and BH_4^- in the layer of borohydride remain their original configuration in pristine $LiBH_4$. Li^+ has a tetrahedral coordination with one AB and three BH_4^- groups. The distances between Li and H in BH_4^- are in the range of 2.023-2.246 Å, analogous to those between Li and its two close hydridic H in AB (2.078 and 2.321 Å). The BH···HN distances between BH_4^- and its nearby AB range from 2.248 to 2.254 Å, whereas the BH···HN distance between the adjacent AB is 2.439 Å. Such a weak BH···HN interaction in AB layers may be ascribed to the reinforced interactions of AB with the nearby H in BH_4^- and also with Li^+. Consequently, the intercalation of AB molecules into the crystal structure of $LiBH_4$ is basically stabilized by the interactions of AB with Li^+ and BH_4^-. There were two steps for dehydrogenation of $2LiBH_4 \cdot AB$, i.e., the first step occurring at 105-135 °C and the second one at 360 °C, which are determined to be the dehydrogenation of pristine AB and $LiBH_4$, respectively. In addition, the suppression of ammonia was achieved compared with pristine AB. It is worth noting that the dehydrogenated products of $2LiBH_4 \cdot AB$ could reabsorb 2.4 wt % H_2 at 400 °C under 82 bar H_2 pressure. In contrast, MBH_4-AB (M = Na, K) do not form new compounds during mechanochemical treatment [76].

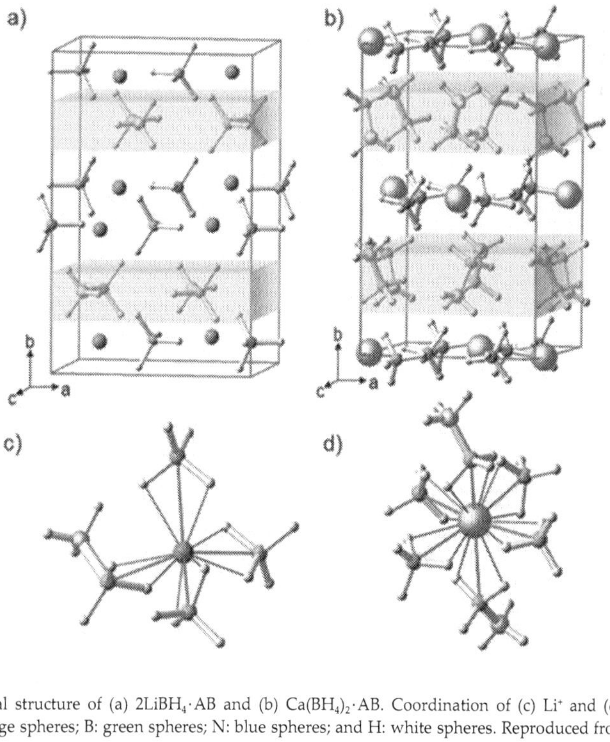

Figure 24. Crystal structure of (a) $2LiBH_4 \cdot AB$ and (b) $Ca(BH_4)_2 \cdot AB$. Coordination of (c) Li^+ and (d) Ca^{2+}. Li: pink spheres; Ca: orange spheres; B: green spheres; N: blue spheres; and H: white spheres. Reproduced from Ref. [75] with permission.

6.2. $LiBH_4 \cdot NH_3BH_3$, $LiBH_4 \cdot AB$

$LiBH_4 \cdot AB$ was prepared by mechanically milling AB with an equivalent $LiBH_4$ [77]. It crystallizes in a monoclinic cell (space group $P2_1$) with lattice parameters of $a = 14.3131$ (11) Å, b= 4.3634 (5) Å, c= 15.3500(13) Å, and β= 90.325(11)°. In $LiBH_4 \cdot AB$, there are $LiBH_4$ columns running along the b axis (Figure 25), which is notably different from $2LiBH_4 \cdot AB$. $LiBH_4$ columns are segregated by the adjacent AB molecules and thus lose the structural similarity to pristine one. In such an arrangement, Li^+ has two different coordination environments, both of which have a trigonal-planar coordination. One is coordinated by two BH_4^- and AB groups and the other by three BH_4^- groups. The unique coordination of Li^+ results in the remarkable variations of the interactions of Li^+ with its adjacent anions. For example, the distances of Li–B are ranging from 2.22 to 2.81 Å. While the distances between Li and its two close hydridic H in AB are 1.78 and 2.07 Å, much smaller than those in $2LiBH_4 \cdot AB$ (2.08 and 2.32 Å) [75], which indicates a stronger interaction between AB and Li^+. The BH···HN distances between AB and the adjacent BH_4^- are in a range of 1.85–2.31 Å, and those between nearby AB in AB layers are 1.86–2.39 Å, which demonstrates the presence of strong dihydrogen bonding throughout $LiBH_4 \cdot AB$. For thermolysis, $LiBH_4 \cdot AB$ first decomposed into $2LiBH_4 \cdot AB$ and AB at about 54 °C and then exhibited a three-step dehydrogenation, with release of a total amount of ~15.7 wt % H_2 upon heating to 450 °C. Interestingly, h-BN was found to form in the post-dehydrogenated product at a moderate temperature of 450 °C (compared with other methods for h-BN preparation), which resulted in the improved rehydrogenation and the following dehydrogenation properties of $LiBH_4 \cdot AB$ sample.

6.3. $Ca(BH_4)_2 \cdot NH_3BH_3$, $Ca(BH_4)_2 \cdot AB$

$Ca(BH_4)_2 \cdot AB$ was prepared by ball milling $Ca(BH_4)_2$ with an equivalent of AB, which had an orthorhombic cell (space group $Aba2$) with lattice parameters of $a = 8.265(1)$ Å, $b = 13.478(2)$ Å, and $c = 8.136(1)$ Å [75]. Each Ca^{2+} has an octahedral coordination, surrounded by two AB molecules with Ca–B distances ranging from 2.899 to 2.923 Å and four BH_4^- groups (Figure 24). Unlike $2LiBH_4 \cdot AB$, the BH···HN distance between neighboring AB in AB layers of $Ca(BH_4)_2 \cdot AB$ is 1.735 Å, which indicates more powerful dihydrogen bonding. Through dihydrogen bonding, AB also interacts with its adjacent BH_4^- with the BH···HN distances of 1.986–2.037 Å. Furthermore, the distances between Ca^{2+} and its two close hydridic H of AB are 2.441 and 2.504 Å, similar to those between Ca^{2+} and H in BH_4^- (2.412–2.477 Å). As a consequence, different from $2LiBH_4 \cdot AB$, AB layers and $Ca(BH_4)_2$ layers are interlaced primarily by dihydrogen bonding in $Ca(BH_4)_2 \cdot AB$. For the case of hydrogen release, it has similar dehydrogenation behaviors with $2LiBH_4 \cdot AB$, i.e., independent dehydrogenation of AB and $Ca(BH_4)_2$ was observed, although the dehydrogenation temperature was slightly reduced to some extent.

6.4. $Mg(BH_4)_2 \cdot 2NH_3BH_3$, $Mg(BH_4)_2 \cdot 2AB$

$Mg(BH_4)_2 \cdot 2AB$ was prepared by ball milling $Mg(BH_4)_2$ with 2 equivalent AB [78,79]. It has an orthorhombic unit cell (space group $P2_12_12_1$) with lattice parameters of a = 14.4135(2) Å, b = 13.2084(2) Å, and c = 5.1118(1) Å. As shown in Figure 26, each Mg^{2+} has a tetrahedral coordi-

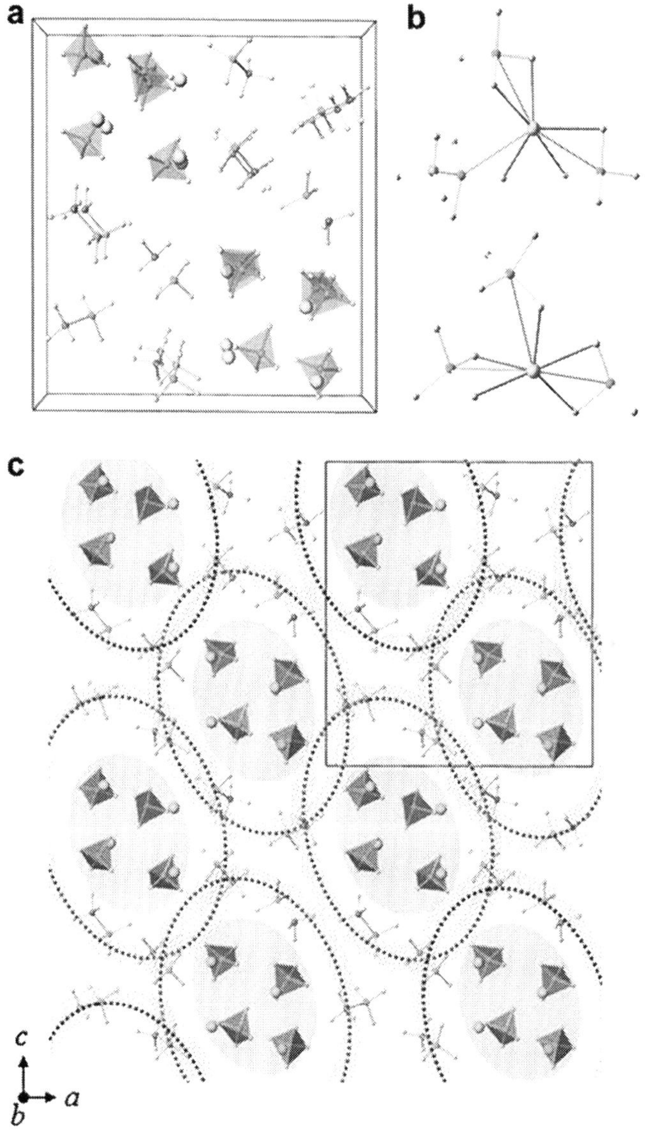

Figure 25. (a) Crystal structure of $LiBH_4 \cdot AB$. (b) Li^+ coordination. (c) Schematic arrangement of $LiBH_4$ and AB components in the crystal structure. Li: yellow spheres; B: green spheres; N: blue spheres; and H: white spheres. Reproduced from Ref. [77] with permission.

nation, surrounded by two BH_4^- groups and two AB groups with Mg-B distances in the range of 2.394–2.500 Å. The distances between Mg^{2+} and its close H in AB range from 2.062 to 2.106 Å, while the distances between Mg^{2+} and its close H in BH_4^- are in the range of 1.987-2.033 Å. Between AB and its nearby BH_4^- groups, there are BH···HN dihydrogen bonds with the distances in the range of 2.210-2.274 Å. $Mg(BH_4)_2{\cdot}2AB$ melted at 48 °C and started to release H_2 at 75 °C [79]. It had a broad dehydrogenation peak at 126 °C with a releasing trace amount of NH_3, B_2H_6, and borazine. The following dehydrogenation in the range of 200-400 °C was observed with two peaks at 297 °C and 340 °C. Volumetric release tests implied that the dehydrogenation behaviors of $Mg(BH_4)_2{\cdot}2AB$ look like a combination of pristine AB in 30-200 °C and $Mg(BH_4)_2$ in 200-400 °C.

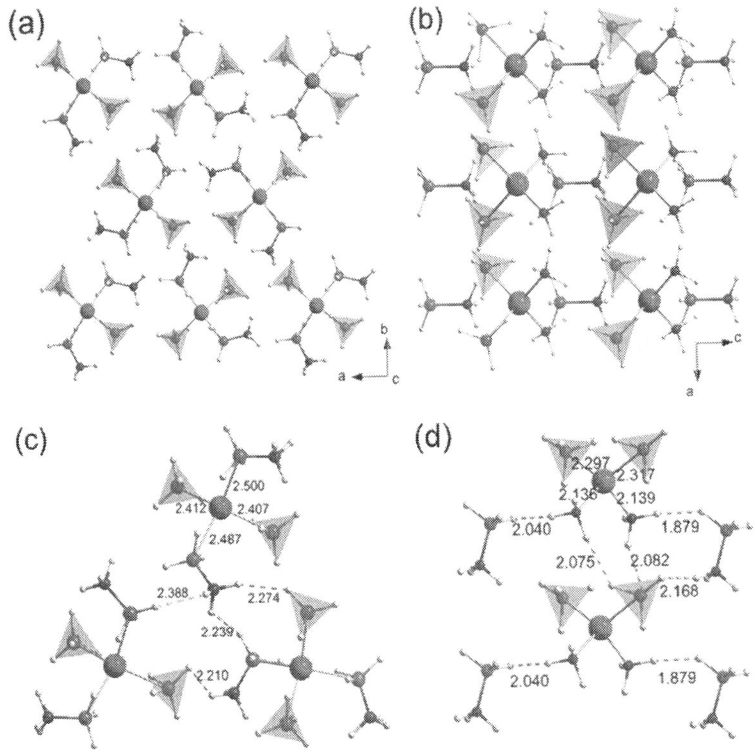

Figure 26. Crystal structure of (a) $Mg(BH_4)_2{\cdot}2AB$ and (b) $Mg(BH_4)_2{\cdot}2NH_3{\cdot}AB$. Coordination of Mg^{2+} in (c) $Mg(BH_4)_2{\cdot}2AB$ and (d) $Mg(BH_4)_2{\cdot}2NH_3{\cdot}AB$. Mg: red spheres; B: pink spheres; N: blue spheres; and H: white spheres. Reproduced from Ref. [78] with permission.

6.5. $Al(BH_4)_3 \cdot NH_3BH_3$, $Al(BH_4)_3 \cdot AB$

$Al(BH_4)_3 \cdot AB$ was synthesized through a reaction between $Al(BH_4)_3$ and an equivalent AB in a closed reactor at room temperature for 3 days, which could be accelerated by ball milling [80]. The resulting white crystals of $Al(BH_4)_3 \cdot AB$ revealed two polymorphs: the low-temperature α-phase and the high-temperature β-phase. Pure α-phase was found only in fresh samples, and it slowly transformed to β-phase at room temperature. Pure β-phase was obtained upon heating α-phase to about 62 °C with 1 °C min^{-1}, and it could not transform back to α-phase even cooling to -173 °C. Both phases crystallize in monoclinic cells with space group of $P2_1/c$ and Cc for α- and β-phase, respectively. α-phase has the lattice parameters of a = 7.8585(2) Å, b = 6.86473(14) Å, c = 15.7136(8) Å, and β = 96.429(4)°, while β-phase of a = 10.8196(8) Å, b = 7.2809(4) Å, c = 11.3260(9) Å, and β = 107.695(8)°. The crystal structures of $Al(BH_4)_3 \cdot AB$ in α-phase and β-phase are shown in Figure 27. In both structures, Al^{3+} is coordinated with one AB and three BH_4^- to form a distorted tetrahedron of AlB4. The interactions of NH···HB between NH_3 and BH_4^- groups were formed in $Al(BH_4)_3 \cdot AB$ to sustain its 3D structure. The distances between Al and B in BH_4^- narrowly range from 2.21 to 2.23 Å, while the distance between Al and B in BH_3 from AB is 2.31 Å. The distances between Al and H in BH_4^- are in the range of 1.65–1.81 Å, while the distances between Al and H in BH_3 range from 1.86 to 1.96 Å. The thermal decomposition of $Al(BH_4)_3 \cdot AB$ at 70 °C gave rise to the release of 2 Eq of H_2 (5 wt%), with a unique intermediate found in [6H + 1H] coordination of Al^{3+}. It is worth noting that the endothermic dehydrogenation of $Al(BH_4)_3 \cdot AB$ (39 kJ mol^{-1}) was achieved.

7. Other derivatives

7.1. $Mg(BH_4)_2 \cdot 2NH_3 \cdot NH_3BH_3$, $Mg(BH_4)_2 \cdot 2NH_3 \cdot AB$

$Mg(BH_4)_2 \cdot 2NH_3 \cdot AB$ was prepared by ball milling $Mg(BH_4)_2 \cdot 2NH_3$ with an equivalent AB [78]. It had a tetragonal cell in space group of $P4bm$ and the lattice parameters of a = 9.4643(8) Å and c = 5.5229(8) Å (Figure 26). Each Mg^{2+} is tetrahedrally coordinated by two BH_4^- with Mg-B distance of 2.30 Å and two NH_3 groups with Mg-N distance of 2.14 Å. NH_3 groups interact with their close BH_4^- and AB groups through the formation of BH···HN dihydrogen bonds(1.821-2.368 Å). Different from $Mg(BH_4)_2 \cdot 2AB$, $Mg(BH_4)_2 \cdot 2NH_3 \cdot AB$ exhibited significant depression of by-products and tremendously improved H_2 release. The onset dehydrogenation was reduced to 75 °C, and about 9.6 wt% H_2 could be released by 170 °C. Upon heating to 400 °C, a total amount of 13.8 wt% H_2 was desorbed.

For dehydrogenation improvement, $Zr(BH_4)_4 \cdot 8NH_3$ was ball milled with four equivalent AB. Although no new compound formed after ball milling and the subsequent heating treatment, it had an onset temperature at about 60 °C for dehydrogenation and could release 7.0 wt% H_2 at 100 °C in 45 min [81].

http://dx.doi.org/10.5772/61048

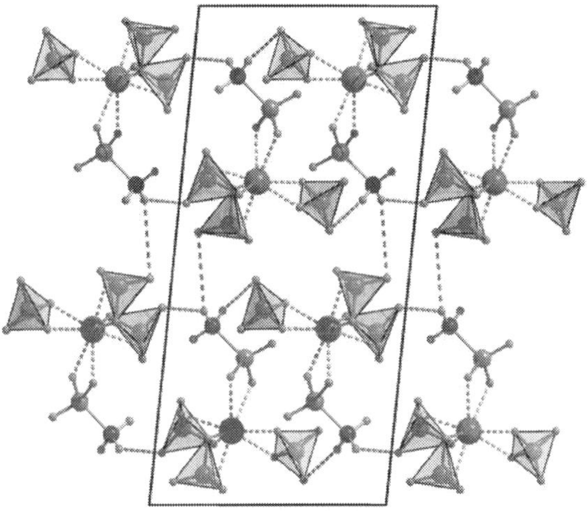

a)

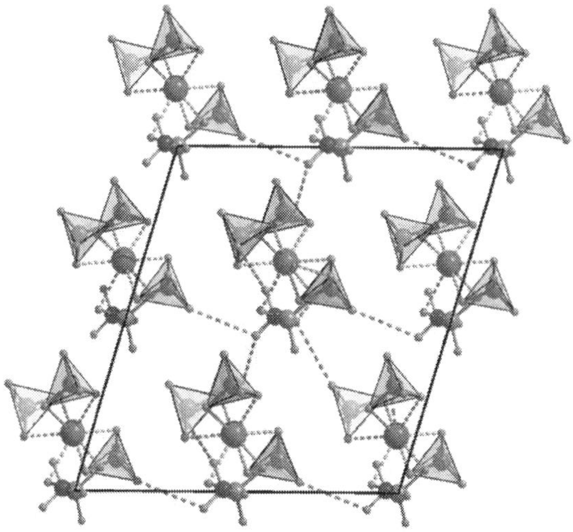

b)

Figure 27. Crystal structure of $Al(BH_4)_3 \cdot AB$: (a) α-phase and (b) β-phase. Reproduced from Ref. [80] with permission.

7.2. $LiCa(NH_2)_3(BH_3)_2$

$LiCa(NH_2)_3(BH_3)_2$ was reported to be prepared by ball milling CaAB with an equivalent $LiNH_2$ [82]. Primary analysis showed that it crystallized in a monoclinic cell with lattice parameters of a = 7.3021 Å, b = 12.5513 Å, c = 5.0595 Å, and β = 120.98°. However, the detailed crystal structure was not given. The onset temperature for dehydrogenation was at about 50 °C and a total amount of 6.8 wt% H_2 could be released exothermically upon heating to 300 °C, without the detectable borazine and diborane.

7.3. Sodium aminodiborane ($NaNH_2(BH_3)_2$)

$NaNH_2(BH_3)_2$ with an infrequent $NH_2(BH_3)_2^-$ anion was first reported by Girolami and coworkers in 2010 [83]. It was synthesized by the reduction of AB with excess Na in THF at room temperature and the following reflux. However, the target compound failed to isolate from the resulting THF adduct. In a very recent paper [84], solvent-free $NaNH_2(BH_3)_2$ was prepared by reacting $NaNH_2$ and two equivalent AB in THF. In a closed reactor, $NaNH_2(BH_3)_2$ could release 2 Eq of H_2 (6.0 wt%) at 271 °C, forming solid products of $NaBH_4$ and highly condensed polyborazylene.

7.4. $LiNH_2BH_3 \cdot LiBH_4$, $LiAB \cdot LiBH_4$

$LiAB \cdot LiBH_4$ was synthesized by ball milling LiAB with an equivalent $LiBH_4$ [85]. It was ascribed to an orthorhombic cell (space group *Pbca*) with lattice parameters of a = 9.2824(18) Å, b = 14.3092(28) Å, and c = 7.6194(12) Å. In $LiAB \cdot LiBH_4$ structure shown in Figure 28, alternative LiAB and $LiBH_4$ layers are stacked along the b axis. A strong interaction is resulting from the formed ionic bonds by Li^+ with $[NH_2BH_3]^-$ and BH_4^-. Li^+ has two tetrahedral coordination environments. One is surrounded by one BH_3 of $[NH_2BH_3]^-$ and three BH_4^-, while the other by one NH_2 of $[NH_2BH_3]^-$, one BH_4^-, and two BH_3 of $[NH_2BH_3]^-$. Furthermore, the BH···HN dihydrogen bondings range from 2.181 to 2.387 Å. However, the detailed dehydrogenation behavior of this compound was not given, although it had a similar process to 2LiAB/$LiBH_4$ reported in detail.

7.5. $M(BH_3NH_2BH_2NH_2BH_3)$, M = Li or Na

Preparation of $M(BH_3NH_2BH_2NH_2BH_3)$ (denoted as M(B3N2), M = Li or Na) was performed through a wet-chemical method in THF, according to the reaction: $MH + 3NH_3BH_3 \rightarrow M(BH_3NH_2BH_2NH_2BH_3) + NH_3 + 2H_2$ [86]. As stable white solids at room temperature, Li(B3N2) crystallizes in a tetragonal cell (space group $P\bar{4}2c$) with lattice parameters of a = 4.02(1) Å and c = 16.95(5) Å, while Na(B3N2) in a hexagonal cell (space group $P6_3/m$) of a = 4.3392(2) Å and c = 17.853(6) Å. Both crystal structures are shown in Figure 29. In Li(B3N2), each layer of Li^+ is linked by $(BH_3NH_2BH_2NH_2BH_3)^-$. Li^+ is tetrahedrally coordinated with four H atoms from terminal BH_3. The shortest distance between Li and H is 2.0 Å, and the average B-N bond length in $(BH_3NH_2BH_2NH_2BH_3)^-$ is about 1.59 Å. The twisty $(BH_3NH_2BH_2NH_2BH_3)^-$ were observed to lie along the c axis in Li(B3N2). It is worth noting that there is no interactions of BH···HN between neighboring $(BH_3NH_2BH_2NH_2BH_3)^-$. For the case of Na(B3N2) structure, there are some similarities to Li(B3N2), i.e., Na^+ layers perpendicular to the c axis separated by twisty $(BH_3NH_2BH_2NH_2BH_3)^-$. The most obvious difference is the fact that Na^+ is coordinated with eight H atoms with the closest Na-H distance of 2.22(9) Å. Upon heating, M(B3N2) decomposed exothermally into MBH_4 and H_2. As an example, Li(B3N2) had a weight loss of 5.0 wt% at 140 °C, contributed to hydrogen release without any ammonia or other impurities.

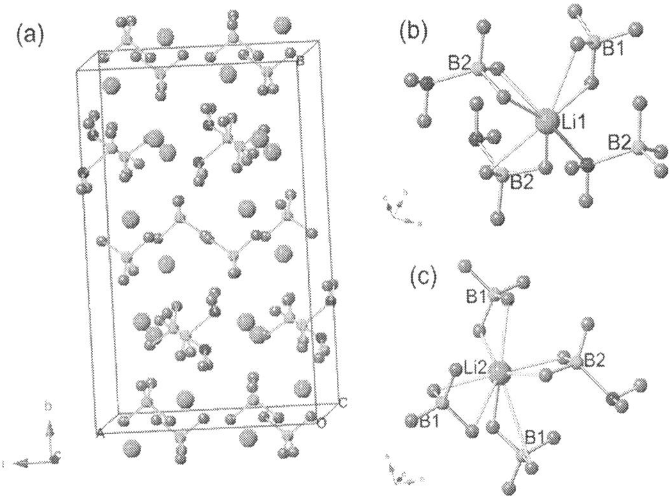

Figure 28. (a) Crystal structure of $LiAB \cdot LiBH_4$ and (b, c) Coordination of Li^+. Li: green spheres; N: blue spheres; B: cyan spheres; and H: pink spheres. Reproduced from Ref. [85] with permission.

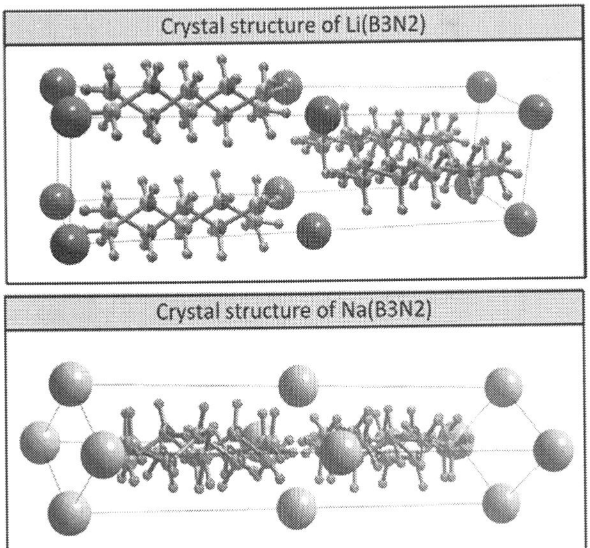

Figure 29. Crystal structure of Li(B3N2) and Na(B3N2): Li: green spheres; Na: yellow spheres; B: red spheres; N: blue spheres; and H: grey spheres. Reproduced from Ref. [86] with permission.

8. Conclusions

In this chapter, metal amidoboranes, metal borohydride-ammonia borane complexes, and metal amidoborane ammoniates as well as their derivatives were summarized with focus on their structures and dehydrogenation properties. In addition, nanoconfinement and nanocatalysis of AB for improving the dehydrogenation were also included. From scientific point of view, some new B–N-containing compounds and their crystal structures were reported, which enriches the chemistry of the scope of B and N elements. For the case of practical applications, the fast dehydrogenation rate at around 60 °C and the stability under 40 °C is one of the most important targets. Due to the difficulties in reversible hydrogenation of AB-related materials, off-board regeneration of spent fuel with high-energy efficiency is another issue to be solved. This would make it vitally important for altering the dehydrogenation thermodynamics from exothermicity to endothermicity. What is exciting is the fact that some cases of endothermic dehydrogenation were observed [60,62,80]. In this regard, it is indispensable for material design. That means that both a large number of experimental studies and the related theoretical calculations are urgently needed in the future. Moreover, catalytic modification should be considered for the dehydrogenation kinetics. In addition, how to eliminate the un-wanted by-products including ammonia, diborane, and borazine in the evolved hydrogen is also a critical issue, which needs special attention to be given because they are detrimental to proton exchange membrane (PEM) fuel cell, especially ammonia even with trace amount. As an example, there is a controversial debate of NaAB dehydrogenation for ammonia contamination [36,49,51]. Therefore, the mechanism for NH_3 formation during dehydrogenation is another task of further study.

Acknowledgements

The related research in this chapter was financially supported by the National Natural Science Foundation of China (51361006, 51401059, 51461010, 51361005, and 51371060), the Guangxi Natural Science Foundation (2014GXNSFAA118043, 2014GXNSFAA118333, 2013GXNSFBA019034, and 2013GXNSFBA019239), and the Guangxi University Research Project (YB2014132). This work was partially sponsored by Guangxi Collaborative Innovation Center of Structure and Property for New Energy Materials.

Author details

Hailiang Chu[1,2*], Shujun Qiu[1,2], Lixian Sun[1,2] and Fen Xu[1,2]

*Address all correspondence to: chuhailiang@guet.edu.cn

1 Guangxi Key Laboratory of Information Materials, Guilin University of Electronic Technology, Guilin, PR China

2 School of Materials Science and Engineering, Guilin University of Electronic Technology, Guilin, China

References

[1] Staubitz A, Robertson APM, Manners I. Ammonia-borane and related compounds as dihydrogen sources. Chemical Reviews 2010;110(7):4079-4124.

[2] Gao L, Li CYV, Yung H, Chan KY. A Functionalized MIL-101(Cr) metal-organic framework for enhanced hydrogen release from ammonia borane at low temperature. Chemical Communications 2013;49(90):10629-10631.

[3] Smythe NC, Gordon JC. Ammonia borane as a hydrogen carrier: dehydrogenation and regeneration. European Journal of Inorganic Chemistry 2010;4:509-521.

[4] Wolf G, Baumann J, Baitalow F, Hoffmann FP. Calorimetric process monitoring of thermal decomposition of B-N-H compounds. Thermochimica Acta 2000;343(1-2): 19-25.

[5] Hu MG, Geanangel RA, Wendlandt WW. The Thermal decomposition of ammonia borane. Thermochimica Acta 1978;23(2): 249-255.

[6] Gutowska A, Li L, Shin YS, Wang CMM, Li XHS, Linehan JC, Smith RS, Kay BD, Schmid B, Shaw W, Gutowski M, Autrey T. Nanoscaffold mediates hydrogen release and the reactivity of ammonia borane. Angewandte Chemie International Edition 2005;44(23):3578-3582.

[7] Feaver A, Sepehri S, Shamberger P, Stowe A, Autrey T, Cao G. Spectroscopic studies of dehydrogenation of ammonia borane in carbon cryogel. Journal of Physical Chemistry B 2007;111(51):7469-7472.

[8] Neiner D, Karkamkar A, Linehan JC, Arey B, Autrey T, Kauzlarich SM. Promotion of hydrogen release from ammonia borane with mechanically activated hexagonal boron nitride. Journal of Physical Chemistry C 2009;113(3):1098-1103.

[9] Bluhm ME, Bradley MG, Butterick R, Kusari U, Sneddon LG. Amineborane-based chemical hydrogen storage: enhanced ammonia borane dehydrogenation in ionic liquids. Journal of the American Chemical Society 2006;128(24):7748-7749.

[10] Zhang JS, Lee JW. Progress and prospects in thermolytic dehydrogenation of ammonia borane for mobile applications. Korean Journal of Chemical Engineering 2012;29(4):421-431.

[11] Denney MC, Pons V, Hebden TJ, Heinekey DM, Goldberg KI. Efficient catalysis of ammonia borane dehydrogenation. Journal of the American Chemical Society 2006;128(37):12048-12049.

[12] Zhang L, Tu Q, Chen X, Liu P. Nano metal catalysts in dehydrogenation of ammonia borane. Progress in Chemistry 2014;26(5):749-755.

[13] Zimmerman PM, Paul A, Zhang ZY, Musgrave CB. The role of free N-heterocyclic carbene (NHC) in the catalytic dehydrogenation of ammonia-borane in the nickel NHC system. Angewandte Chemie International Edition 2009;48(12):2201-2205.

[14] Zhao J, Shi J, Zhang X, Cheng F, Liang J, Tao Z, Chen J. A soft hydrogen storage material: poly(methyl acrylate)-confined ammonia borane with controllable dehydrogenation. Advanced Materials 2010;22(3):394-397.

[15] Chua YS, Chen P, Wu G, Xiong Z. Development of amidoboranes for hydrogen storage. Chemical Communications 2011;47(18):5116-5129.

[16] Wang K, Zhang J, Man T, Wu M, Chen C. Recent process and development of metal aminoborane. Chemistry—An Asian Journal 2013;8(11):1076-1089.

[17] Wolstenholme DJ, Titah JT, Che FN, Traboulsee KT, Flogeras J, McGrady GS. Homopolar dihydrogen bonding in alkali-metal amidoboranes and its implications for hydrogen storage. Journal of the American Chemical Society 2011;133(41):16598-16604.

[18] Luedtke AT, Autrey T. Hydrogen release studies of alkali metal amidoboranes. Inorganic Chemistry 2010;49(8):3905-3910.

[19] Jepsen LH, Ley MB, Lee YS, Cho YW, Dornheim M, Jensen JO, Filinchuk Y, Jørgensen JE, Besenbacher F, Jensen TR. Boron-nitrogen based hydrides and reactive composites for hydrogen storage. Materials Today 2014;17(3):129-135.

[20] Moussa G, Moury R, Demirci UB, Şener T, Miele P. Boron-based hydrides for chemical hydrogen storage. International Journal of Energy Research 2013;37(8):825-842.

[21] Furukawa H, Cordova KE, O'Keeffe M, Yaghi OM. The chemistry and applications of metal-organic frameworks. Science 2013;341(6149):1230444.

[22] Zhou HC, Long JR, Yaghi OM. Introduction to metal-organic frameworks. Chemical Reviews 2012;112(2):673-674.

[23] Li Z, Zhu G, Lu G, Qiu S, Yao X. Ammonia borane confined by a metal-organic framework for chemical hydrogen storage: enhancing kinetics and eliminating ammonia. Journal of the American Chemical Society 2010;132(5):1490-1491.

[24] Gadipelli S, Ford J, Zhou W, Wu H, Udovic TJ, Yildirim T. Nanoconfinement and catalytic dehydrogenation of ammonia borane by magnesium-metal-organic-framework-74. Chemistry—A European Journal 2011;17(22):6043-6047.

[25] Si X, Sun L, Xu F, Jiao C, Li F, Liu S, Zhang J, Song L, Jiang C, Wang S, Liu Y, Sawada Y. Improved hydrogen desorption properties of ammonia borane by Ni-modified metal-organic frameworks. International Journal of Hydrogen Energy 2011;36(11):6698-6704.

[26] Cheng F, Ma H, Li Y, Chen J. $Ni_{1-x}Pt_x$ (x = 0-0.12) hollow spheres as catalysts for hydrogen generation from ammonia borane. Inorganic Chemistry 2007;46(3):788-794.

[27] He T, Xiong Z, Wu G, Chu H, Wu C, Zhang T, Chen P. Nanosized Co- and Ni-catalyzed ammonia borane for hydrogen storage. Chemistry of Materials 2009;21(11): 2315-2318.

[28] He T, Wang J, Wu G, Kim H, Proffen T, Wu A, Li W, Liu T, Xiong Z, Wu C, Chu H, Guo J, Autrey T, Zhang T, Chen P. Growth of Crystalline polyaminoborane through catalytic dehydrogenation of ammonia borane on FeB nanoalloy. Chemistry—A European Journal 2010;16(43):12814-12817.

[29] He T, Wang J, Liu T, Wu G, Xiong Z, Yin J, Chu H, Zhang T, Chen P. Quasi in situ Mössbauer and XAS studies on FeB nanoalloy for heterogeneous catalytic dehydrogenation of ammonia borane. Catalysis Today 2011;170(1):69-75.

[30] Li L, Yao X, Sun C, Du A, Cheng L, Zhu Z, Yu C, Zou J, Smith SC, Wang P, Cheng H, Frost RL, Lu G. Lithium-catalyzed dehydrogenation of ammonia borane within mesoporous carbon framework for chemical hydrogen storage. Advanced Functional Materials 2009;19(2):265-271.

[31] Li Y, Song P, Zheng J, Li X. Promoted H_2 generation from NH_3BH_3 thermal dehydrogenation catalyzed by metal-organic framework based catalysts. Chemistry—A European Journal 2010;16(35):10887-10892.

[32] Li S, Guo Y, Sun W, Sun D, Yu X. Platinum nanoparticle functionalized CNTs as nanoscaffolds and catalysts to enhance the dehydrogenation of ammonia-borane. Journal of Physical Chemistry C 2010;114(49):21885-21890.

[33] Xin G, Yang J, Li W, Zheng J, Li X. Catalytic thermal decomposition of ammonia-borane by well-dispersed metal nanoparticles on mesoporous substrates prepared by magnetron sputtering. European Journal of Inorganic Chemistry 2012;34: 5722-5728.

[34] Schlesinger HI, Burg AB. Hydrides of Boron. VIII. The structure of the diammoniate of diborane and its relation to the structure of diborane. Journal of the American Chemical Society 1938;60(2):290-299.

[35] Myers AG, Yang BH, David KJ. Lithium amidotrihydroborate, a powerful new reductant. Transformation of tertiary amides to primary alcohols. Tetrahedron Letters 1996;37(21):3623-3626.

[36] Xiong Z, Yong CK, Wu G, Chen P, Shaw W, Karkamkar A, Autrey T, Jones MO, Johnson SR, Edwards PP, David WIF. High-capacity hydrogen storage in lithium and sodium amidoboranes. Nature Materials 2008;7:138-141.

[37] Kang X, Fang Z, Kong L, Cheng H, Yao X, Lu G, Wang P. Ammonia borane destabilized by lithium hydride: an advanced on-board hydrogen storage material. Advanced Materials 2008;20(14):2756-2759.

[38] Wu H, Zhou W, Yildirim T. Alkali and alkaline-earth metal amidoboranes: structure, crystal chemistry, and hydrogen storage properties. Journal of the American Chemical Society 2008;130(44):14834-14839.

[39] Diyabalanage HVK, Shrestha RP, Semelsberger TA, Scott BL, Bowden ME, Davis BL, Burrell AK. Calcium amidotrihydroborate: a hydrogen storage material. Angewandte Chemie International Edition 2007;46(47):8995-8997.

[40] Diyabalanage HVK, Nakagawa T, Shrestha RP, Semelsberger TA, Davis BL, Scott BL, Burrell AK, David WIF, Ryan KR, Jones MO, Edwards PP. Potassium(I) amidotrihydroborate: structure and hydrogen release. Journal of the American Chemical Society 2010;132(34):11836-11837.

[41] Zhang Q, Tang C, Fang C, Fang F, Sun D, Ouyang L, Zhu M. Synthesis, crystal structure, and thermal decomposition of strontium amidoborane. Journal of Physical Chemistry C 2010;114(3):1709-1714.

[42] Genova RV, Fijalkowski KJ, Budzianowski A, Grochala W. Towards $Y(NH_2BH_3)_3$: probing hydrogen storage properties of YX_3/MNH_2BH_3 (X = F, Cl; M= Li, Na) and $YH_{x\sim3}/NH_3BH_3$ composites. Journal of Alloys and Compounds 2010;499(2):144-148.

[43] Lu J, Fang Z. Dehydrogenation of a combined $LiAlH_4/LiNH_2$ system. Journal of Physical Chemistry B 2005;109(44): 20830-20834.

[44] Chen P, Xiong Z, Luo J, Lin J, Tan K. Interaction between lithium amide and lithium hydride. Journal of Physical Chemistry B 2003;107(39):10967-10970.

[45] Xiong Z, Wu G, Hu J, Chen P. Ternary Imides for hydrogen storage. Advanced Materials 2004;16(17):1522-1525.

[46] Hughes EW. The crystal structure of ammonia-borane, H_3NBH_3. Journal of the American Chemical Society 1956;78(2):502-503.

[47] Bowden ME, Gainsford GJ, Robinson WT. Room-temperature structure of ammonia borane. Australian Journal of Chemistry 2007;60(3):149-153.

[48] Xiong Z, Chua YS, Wu G, Xu W, Chen P, Shaw W, Karkamkar A, Linehan J, Smurthwaite T, Autrey T. Interaction of lithium hydride and ammonia borane in THF. Chemical Communications 2008;43:5595-5597.

[49] Xia G, Chen J, Sun W, Tan Y, Guo Z, Liu H, Yu X. Well-dispersed lithium amidoborane nanoparticles through nanoreactor engineering for improved hydrogen release. Nanoscale 2014;6(21):12333-12339.

[50] Xiong Z, Wu G, Chua YS, Hu J, He T, Xu W, Chen P. Synthesis of sodium amidoborane ($NaNH_2BH_3$) for hydrogen production. Energy and Environmental Science 2008;1(3):360-363.

[51] Shimoda K, Zhang Y, Ichikawa T, Miyaoka H, Kojima Y. Solid state NMR study on the thermal decomposition pathway of sodium amidoborane $NaNH_2BH_3$. Journal of Materials Chemistry 2011;21(8):2609-2615.

[52] Fijałkowski KJ, Grochala W. Substantial emission of NH_3 during thermal decomposition of sodium amidoborane, $NaNH_2BH_3$. Journal of Materials Chemistry 2009;19(14):2043-2050.

[53] Wu C, Wu G, Xiong Z, David WIF, Ryan KR, Jones MO, Edwards PP, Chu H, Chen P. Stepwise phase transition in the formation of lithium amidoborane. Inorganic Chemistry 2010;49(9):4319-4323.

[54] Wu C, Wu G, Xiong Z, Han X, Chu H, He T, Chen P. $LiNH_2BH_3 \cdot NH_3BH_3$: structure and hydrogen storage properties. Chemistry of Materials 2010;22(1):3-5.

[55] Luo J, Kang X, Wang P. Synthesis, formation mechanism, and dehydrogenation properties of the long-sought $Mg(NH_2BH_3)_2$ compound. Energy and Environmental Science 2013;6(3):1018-1025.

[56] Spielmann J, Jansen G, Bandmann H, Harder S. Calcium amidoborane hydrogen storage materials: crystal structures of decomposition products. Angewandte Chemie International Edition 2008;47(33):6290-6295.

[57] He T, Wang J, Chen Z, Wu A, Wu G, Yin J, Chu H, Xiong Z, Zhang T, Chen P. Metathesis of alkali-metal amidoborane and $FeCl_3$ in THF. Journal of Materials Chemistry 2012;22(15):7478-7483.

[58] Fijalkowski KJ, Genova RV, Filinchuk Y, Budzianowski A, Derzsi M, Jaroń T, Leszczynski PJ, Grochala W. $Na[Li(NH_2BH_3)_2]$—the first mixed-cation amidoborane with unusual crystal structure. Dalton Transactions 2011;40(17):4407-4413.

[59] Li W, Miao L, Scheicher RH, Xiong Z, Wu G, Araújo CM, Blomqvist A, Ahuja R, Feng Y, Chen P. Li-Na ternary amidoborane for hydrogen storage: experimental and first-principles study. Dalton Transactions 2012;41(16):4754-4764.

[60] Kang X, Luo J, Zhang Q, Wang P. Combined formation and decomposition of dual-metal amidoborane $NaMg(NH_2BH_3)_3$ for high-performance hydrogen storage. Dalton Transactions 2011;40(15):3799-3801.

[61] Wu H, Zhou W, Pinkerton FE, Meyer MS, Yao QR, Gadipelli S, Udovic TJ, Yildirim T, Rush JJ. Sodium magnesium amidoborane: the first mixed-metal amidoborane. Chemical Communications 2011;47(14):4102-4104.

[62] Chua YS, Li W, Wu G, Xiong Z, Chen P. From exothermic to endothermic dehydrogenation—interaction of monoammoniate of magnesium amidoborane and metal hydrides. Chemistry of Materials 2012;24(18):3574-3581.

[63] Xia G, Tan Y, Chen X, Guo Z, Liu H, Yu X. Mixed-metal (Li, Al) amidoborane: synthesis and enhanced hydrogen storage properties. Journal of Materials Chemistry A 2013;1(5):1810-1820.

[64] Nakagawa Y, Ikarashi Y, Isobe S, Hino S, Ohnuki S. Ammonia borane-metal alanate composites: hydrogen desorption properties and decomposition processes. RSC Advances 2014;4(40):20626-20631.

[65] Sorensen RZ, Hummelshoj JS, Klerke A, Reves JB, Vegge T, Norskov JK, Christensen CH. Indirect, reversible high-density hydrogen storage in compact metal ammine salts. Journal of the American Chemical Society 2008;130(27):8660-8668.

[66] Christensen CH, Sørensen RZ, Johannessen T, Quaade UJ, Honkala K, Elmøe TD, Køhler R, Nørskov JK. Metal ammine complexes for hydrogen storage. Journal of Materials Chemistry 2005;15(38):4106-4108.

[67] Lin R, Liu Y, Gao M, Wang J, Ge H, Pan H. Investigation on performances of the novel ammonia-based hydrogen storage material $CaCl_2$. Journal of Inorganic Materials 2008;23(5):1059-1063.

[68] Soloveichik G, Her JH, Stephens PW, Gao Y, Rijssenbeek J, Andrus M, Zhao JC. Ammine magnesium borohydride complex as a new material for hydrogen storage: structure and properties of $Mg(BH_4)_2 \cdot 2NH_3$. Inorganic Chemistry 2008;47(10): 4290-4298.

[69] Chu H, Wu G, Xiong Z, Guo J, He T, Chen P. Structure and hydrogen storage properties of calcium borohydride diammoniate. Chemistry of Materials 2010;22(21): 6021-6028.

[70] Graham KR, Kemmitt T, Bowden ME. High capacity hydrogen storage in a hybrid ammonia borane-lithium amide material. Energy and Environmental Science 2009;2(6):706-710.

[71] Xia G, Yu X, Guo Y, Wu Z, Yang C, Liu H, Dou S. Amminelithium amidoborane $[Li(NH_3)NH_2BH_3]$: a new coordination compound with favorable dehydrogenation characteristics. Chemistry—A European Journal 2010;16(12):3763-3769.

[72] Chua YS, Wu G, Xiong Z, He T, Chen P. Calcium amidoborane ammoniate-synthesis, structure, and hydrogen storage properties. Chemistry of Materials 2009;21(20): 4899-4904.

[73] Chua YS, Wu H, Zhou W, Udovic TJ, Wu G, Xiong Z, Wong MW, Chen P. Monoammoniate of calcium amidoborane: synthesis, structure, and hydrogen-storage properties. Inorganic Chemistry 2012;51(3):1599-1603.

[74] Chua YS, Wu G, Xiong Z, Karkamkar A, Guo J, Jian MX, Wong MW, Autrey T, Chen P. Synthesis, structure and dehydrogenation of magnesium amidoborane monoammoniate. Chemical Communications 2010;46(31):5752-5754.

[75] Wu H, Zhou W, Pinkerton FE, Meyer MS, Srinivas G, Yildirim T, Udovic TJ, Rush JJ. A new family of metal borohydride ammonia borane complexes: synthesis, structures, and hydrogen storage properties. Journal of Materials Chemistry 2010;20(31): 6550-6556.

[76] Jepsen LH, Skibsted J, Jensen TR. Investigations of the thermal decomposition of MBH_4-$2NH_3BH_3$, M = Na, K. Journal of Alloys and Compounds 2013;580(S1):S287-S291.

[77] Luo J, Wu H, Zhou W, Kang X, Fang Z, Wang P. $LiBH_4 \cdot NH_3BH_3$: a new lithium borohydride ammonia borane compound with a novel structure and favorable hydrogen storage properties. International Journal of Hydrogen Energy 2012;37(14): 10750-10757.

[78] Chen X, Yuan F, Gu Q, Yu X. Synthesis, structures and hydrogen storage properties of two new H-enriched compounds: $Mg(BH_4)_2(NH_3BH_3)_2$ and $Mg(BH_4)_2 \cdot (NH_3)_2(NH_3BH_3)$. Dalton Transactions 2013;42(40):14365-14368.

[79] Jepsen LH, Ban V, Møller KT, Lee YS, Cho YW, Besenbacher F, Filinchuk Y, Skibsted J, Jensen TR. Synthesis, crystal structure, thermal decomposition, and ^{11}B MAS NMR characterization of $Mg(BH_4)_2(NH_3BH_3)_2$. Journal of Physical Chemistry C 2014;118(23):12141-12153.

[80] Dovgaliuk I, Duff CSL, Robeyns K, Devillers M, Filinchuk Y. Mild dehydrogenation of ammonia borane complexed with aluminum borohydride. Chemistry of Materials 2015;27(3):768-777.

[81] Huang J, Tan Y, Gu Q, Ouyang L, Yu X, Zhu M. Ammonia borane modified zirconium borohydride octaammoniate with enhanced dehydrogenation properties. Journal of Materials Chemistry A 2015;3(10):5299-5304.

[82] Liu B, Chua YS, Wu G, Xiong Z, Chen P. Synthesis and dehydrogenation of $LiCa(NH_2)_3(BH_3)_2$. International Journal of Hydrogen Energy 2012;37(11):9076-9081.

[83] Daly SR, Bellott BJ, Kim DY, Girolami GS. Synthesis of the long-sought unsubstituted aminodiboranate $Na(H_3B\text{-}NH_2\text{-}BH_3)$ and its N-Alkyl analogs. Journal of the American Chemical Society 2010;132(21):7254-7255.

[84] Chen W, Huang Z, Wu G, Chen P. New synthetic procedure for $NaNH_2(BH_3)_2$ and evaluation of its hydrogen storage properties. Science China Chemistry 2015;58(1): 169-173.

[85] Chen J, He T, Wu G, Xiong Z, Chen P. Synthesis and hydrogen storage properties of lithium borohydride amidoborane complex. International Journal of Hydrogen Energy 2013;38(25):10944-10949.

[86] Fijalkowski KJ, Jaroń T, Leszczyński PJ, Magos-Palasyuk E, Palasyuk T, Cyrańskic MK, Grochala W. $M(BH_3NH_2BH_2NH_2BH_3)$—the missing link in the mechanism of the thermal decomposition of light alkali metal amidoboranes. Physical Chemistry Chemical Physics 2014;16(42):23340-23346.

Chapter 4

Modeling of Hydrogen Absorption/Desorption in a Metal Hydride Bed Reactor — A Theoretical Study

Olaitan Akanji and Andrei Kolesnikov

Additional information is available at the end of the chapter

http://dx.doi.org/10.5772/61625

Abstract

Hydrogen has been considered as an alternative source of fuel to the fossil fuel in future, most especially, for mobile applications. However, a requirement is the safe, efficient and compact on- board storage of hydrogen. Reversible storage in metal hydride is promising, but adequate knowledge of materials system fulfills all requirements regarding hydrogen content is a major drawback, release temperature, and reversibility simultaneously. Hydrogen absorption- desorption in a metal hydride bed reactor can be modeled using different software such as FLUENT, CFD-ACE, and COMSOL Multiphysics. This book chapter will focus on the use of software COMSOL Multiphysics to simulate the diffusion and heating of hydrogen and metal hydride powder in both radial and axial directions. The model consists of system of partial differential equations (PDE) describing two-dimensional heat and mass transfer of hydrogen in a porous matrix. The influence of the operating parameters Temperature, Pressure, Concentration, Permeability and Thermal Conductivity on the rate of absorption-desorption of hydrogen in metal hydride will be fully discussed. The simulation results obtained could be applied to the on - board hydrogen storage technology, in particular for the hydrogen supply of a fuel cell for powering of a hydrogen fuel cell vehicle.

Keywords: Hydrogen, Storage, Metal hydride, Absorption, Desorption

1. Introduction

The excessive demand of Energy across the globe has risen sharply over the last decade. Today, fossil- fuel- based sources of energy are being depleted at a fast rate because of the ever increasing energy demand and consumption. In addition, fossil- fuels are contributing to both greenhouse gas emissions and global warming [1]. Due to these hazardous effects, there should be an urgent need for alternative and cleaner fuels which will be toxin free and environmentally friendly. Hydrogen energy seems to be the best alternative for the future to replace the fossil fuel and it already has several applications such as in heat pumps and automotive industries. One of the main problems in large usage of hydrogen in automotive industry is the storage problem. The conventional hydrogen storage methods in use for hydrogen gases are gas compression and liquefaction. These storage methods are impractical since the former requires very heavy gas tank and the latter is too expensive to be implement in public vehicles. Storing hydrogen in metal hydride beds as a chemical compound appears to be method of hydrogen storage in near future, and as this has received much interest recently [2]. There are several important challenges to overcome before hydrogen can become a viable fuel. Only 1% of the hydrogen is available as molecular hydrogen gas while the majority is present in the form of water or hydrocarbons. The large- scale production of hydrogen from water or chemical compounds in an efficient and clean method remains a challenge. Another challenge is storage volume. Although hydrogen has high energy content by weight (three times that of gasoline), its energy content by volume is only one-tenth that of gasoline. The process is major storage volume problem in automotive and other mobile systems with stringent volume constraint. A third challenge is hydrogen transport to filling stations. Because of the low density of hydrogen, finding an economical way of transporting large amounts of hydrogen to various locations is quiet illusive. The fourth challenge is the use of hydrogen as a fuel in producing clean energy that would justify its wide acceptance in the market [1]. Hydrogen can be used to produce energy either by burning with oxygen in air (e.g. internal combustion engine) or by electrochemical reaction with oxygen using fuel cells. Fuel cells are both energy efficient and virtually pollution free compared to internal combustion engines. The choice of energy production method is dictated mainly by cost and application. A promising alternative method of storing hydrogen in a fuel cell is solid-metal hydride storage utilizing metal hydride to absorb/desorbs hydrogen at a relatively low pressure(<20-bar), which offers safety and cost advantage, but the obvious disadvantage of weight for hydrogen onboard storage. Efficient release of hydrogen gas in the metal hydride reactor should meet the need for fast load variation in order to build and satisfy such hydrogen systems. The physics of the transport process coupled with reaction kinetics are very important such as hydrogen mass flow in the hydride bed, heat transfer within the bed and local hydrogen absorption rate. - [3]. Several mathematical models for analyzing hydrogen absorption/desorption in metal hydride beds were presented in the literature:

Jemni and Nasrallah [3] presented a model for the two-dimensional transient heat and mass transfer within a cylindrical reactor. The influence of some parameters (reactor radius, temperature and inlet pressure) on the dynamic reactor performance was determined.

Aldas et al. [4] extended the mathematical model of jemni and Nasrallah to three- dimensions demonstrating that hydrogen flow significantly influences the temperature profile in the system

Mayer et al. [5] - developed a one- dimensional model which showed that heat and mass transfer are key factors affecting reaction rate in the reactor.

http://dx.doi.org/10.5772/61625

Nakawaga et al.[6]- predicted the transient heat and mass transfer phenomenon through the hydride bed by using a two-dimensional mathematical model with hydriding and dehydriding kinetics.

2. Methodology

Those equations, which represent the mathematical model of hydrogen absorption/desorption in metal hydride, are stated here, and implemented in COMSOL MULTIPHYSICS software to validate the experiment performed by Jemni et al. [9] and the following parameters were determined:

- Temperature variation during charging and discharging of hydrogen in metal hydride reactor
- Variation in the density of metal hydride in the reactor during hydrogen charging and discharging
- Variation in the pressure of hydrogen in metal hydride reactor during charging and discharging
- Variation in the flow rate of the gas at the outlet of the metal hydride reactor

2.1. Mass balance

The mass conservation equation of the solid metal hydride is presented as follows:

$$(1-\varepsilon)\frac{\partial \rho_s}{\partial t} = m \tag{1}$$

In Equation (1), the left side represents the accumulation/dissipation term for mass of metal hydride, and the term on the right side represents mass of metal hydride release during adsorption/desorption.

Mass balance equation for hydrogen gas is given by equation (2):-

$$\varepsilon\frac{\partial \rho_g}{\partial t} + \nabla\left(\rho_g u_g\right) = m \tag{2}$$

In Equation (2), the term - $\varepsilon\frac{\partial \rho_g}{\partial t}$ - represents accumulation/dissipation term for hydrogen gas release during adsorption/desorption and the second term on the left side represents convection transfer. The term on the right side represents amount of hydrogen gas release during absorption/desorption.

$$u_g = \frac{K}{\mu_g} \nabla P_g \tag{3}$$

Equation (3) is called Darcy equation, and the velocity of hydrogen gas is calculated from this equation. The equation term on the left side represents the velocity of the gas and the right-hand side of the equation represents Darcy expression.

$$m = C_d \exp\left(-\frac{E_d}{R_g T}\right)\frac{P_g - P_{eq}}{P_{eq}}\rho_s \tag{4}$$

The hydrogen release/consumption due to absorption/desorption is given by Equation (4).The term on the left side represents the amount of hydrogen gas released during absorption/desorption and the equation term on the right side represents the kinetic of absorption/desorption of hydrogen.

2.2. Momentum balance and equilibrium pressure

Jemni et al. [3],- proposed the following equilibrium pressure equation

$$\ln\left(P_{eq}\right) = \frac{\Delta H}{RT} - \frac{\Delta S}{R} + \left(\varphi - \phi_0\right)\tan\left[\pi\left(\frac{C}{C_m} - \frac{1}{2}\right)\right] + \frac{\beta}{2} \tag{5}$$

In Equation (5), the term on the left side represents equilibrium pressure. Vant Hoff derived the right- hand side expression, the second term to the right is known as plateau value, and the last term to the right is called hysteresis value.

$$\frac{\partial\left(\rho_g \varepsilon\right)}{\partial t} + \nabla\left(-\rho_g \frac{K}{\mu_g}\nabla P_g\right) = 0 \tag{6}$$

Equation (6), - represents the momentum balance, the first term to the left represents change in momentum due to non-stationary equilibrium and the second term to the left represents momentum change due to absorption inside the porous medium.

$$K = C_k \cdot d_p^2 \left(\frac{\varepsilon}{1-\varepsilon}\right)^2$$

Permeability K and porosity are related Equation (6) above, where d_p is metal hydride particle diameter and constant Ck=2.37x10^{-3}.

2.3. Energy balance

Energy balance is written in assumption that porous metal hydride bed and hydrogen gas are having the same temperature.

Energy balance equation (7) describes temperature evolution of the hydrogen porous bed system-:

$$\frac{d}{dt}\left(\varepsilon \rho_g C_{pg} T + (1-\varepsilon)\rho_s C_{ps} T - \varepsilon \frac{\rho_g RT}{M_{H_2}}\right) + \nabla\left(\rho_g C_{pg} uT - \lambda_e \nabla T\right) + (1-\varepsilon)(-\Delta H)\rho_s m = 0 \tag{7}$$

The first term in Equation (7) to the left represents unsteady heat change in the system, and second term to the left represents heat change in the system due to conduction and third term to the left represents heat release/consumption due to chemical reaction in absorption-desorption processes.

Effective thermal conductivity of metal hydride bed is expressed as:

$$\lambda_e = \varepsilon \lambda_g + (1-\varepsilon)\lambda_s \tag{8}$$

2.4. Conclusion

The dynamic model of the process incorporates coupling between hydrogen adsorption/desorption, flow in porous media, and heat transfer in porous media and gas. The model also assumes thermal equilibrium between metal hydride (solid phase) and hydrogen (gas phase). The equations are implemented in computational software COMSOL 4.0a to solve for dependent variables such as temperature, metal hydride density, gas pressure, and velocity of the gas.

2.5. Building the geometry

The geometry for the MHHST was presented as a cylindrical reactor where hydrogen is adsorbed/desorbed in/out of metal-hydride powder stored in the reactor. Because of axial symmetry, 2D model was used.

In COMSOL software, the geometry was built in the form of rectangles, and final geometry was developed by the Boolean operations: forming union of three different rectangles and then finding the Boolean differences.

One rectangle represents domain for hydrogen, the second rectangle represents the domain for metal-hydride that will absorb/desorbs hydrogen, and the third rectangle represents the domain that will supply heat to the system. The material used in for the tank walls is copper due to its high conductivity. Copper walls were built as an envelope surrounding the metal hydride rectangles. The contour line of the tank was built from certain geometrical points. The points have different coordinates listed in the Table 1. The geometry of the tank is shown in Figure 1. It can be seen that the tank has two radial ribs/fins for better thermal management.

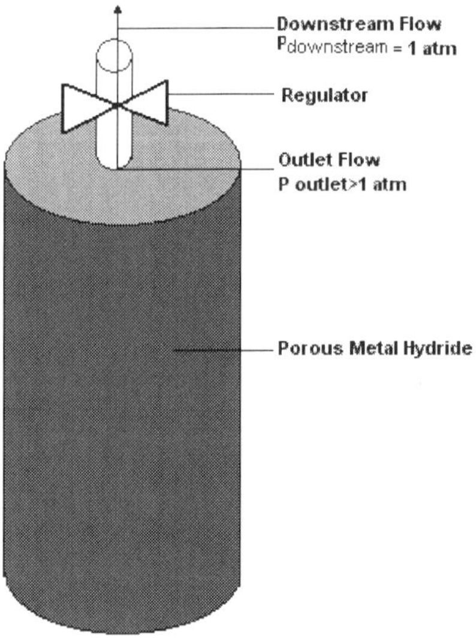

Figure 1. Schematic of Metal hydride reactor.

R- coordinates	Z- coordinates
0.025	0
0.025	0.05
0.025	0.075
0.025	0.15
0.025	0.175
0.025	0.24
0.05	0.05
0.05	0.075

R- coordinates	Z- coordinates
0.05	0.15
0.05	0.175

Table 1. Geometry Coordinates

2.6. Computational mesh

Figure 2 shows the mesh of the MHHST and different boundaries. The portion of the MHHST is taken as rectangular shape that comprises two computational domains (1 and 2). Domain 1 represents metal hydride insulated with copper body (Domain 2), which serves as a conductor to supply/remove heat from metal hydride bed. The heat is supplied to metal hydride (or removed) through the copper walls.

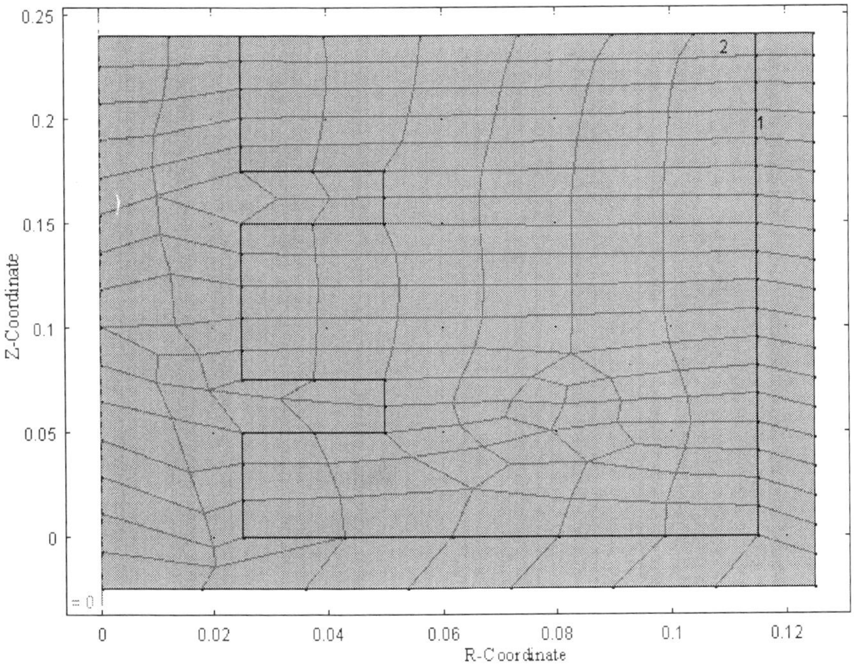

Figure 2. Computational Mesh.

2.7. Boundary conditions

The boundary conditions are described in Figure 3.

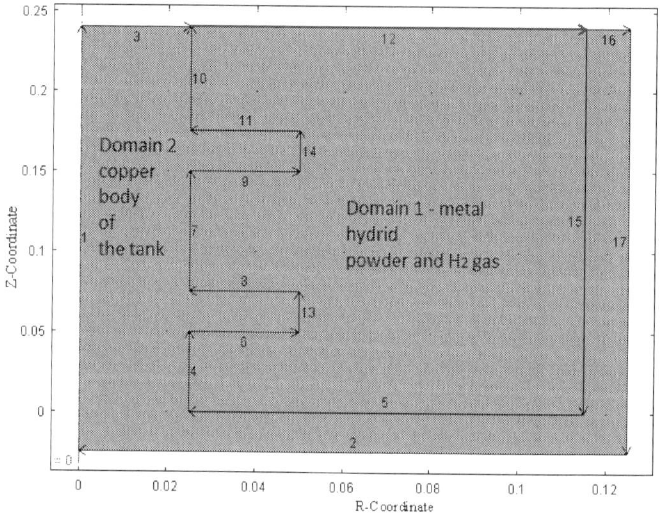

Figure 3. Boundaries within MHHST during hydrogen absorption and desorption

Figure 3. shows different boundaries within MHHST. Each boundary

Illustrates the heat and mass transfer of hydrogen during absorption and desorption.

The Following types of boundary conditions were used.

For modeling of heat transfer during absorption:

a. inlet/outlet –Boundary 12

b. wall – Boundary 9, 14...

c. flux – Boundary 9...

d. isolation – Boundary 9...

For modeling of momentum transfer during absorption:

a. inlet/outlet

b. wall

c. flux

d. isolation

For modeling of hydrogen diffusion during absorption:

a. inlet/outlet Pressure P = 40_bar_(absorption) or 10_bar_(desorption)

b. wall - $C = C_{wall}$ = 293_K

http://dx.doi.org/10.5772/61625

c. flux Q=...h.(T_{ext-} T) mol/m^2

d. isolation - dC/dz= 0

For modeling of heat transfer during desorption:

a. inlet– outlet – Boundary 12

b. wall–Boundary 9,Boundary 14, Boundary 16, T=293_K

c. flux – boundary 9 Q=...h.(T_{ext}- T) W/m^2

d. Isolation – Boundary 9,Boundary 10, Boundary11–. dT/dz = 0

For modeling of momentum transfer during desorption:

a. inlet/outlet

b. Wall

c. Flux

d. Isolation

For modeling of hydrogen released during desorption:

a. inlet/outlet Pressure P =8_bar

b. wall, C=C_{wall} = rho_sat/Mg (mol/m^3)

c. flux Q= h.(T_{ext}-T) (mol/m3)/m^2

d. isolation, dC/dz=0

2.8. Initial conditions

At the beginning of the absorption, the system is at 293_K and the hydrogen is injected at the top of the system (Boundary 12) with the pressure P= 40_bar, then the hydride absorbs and the exothermic reaction raises the temperature.

At the beginning of the desorption, the system is at 333_K and the hydrogen is released from the MHHST through the outlet of the system (Boundary 12) with the pressure P= 10_bar. The temperature decreases slowly within the tank, due to the endothermic reaction, then hydrogen desorbs slowly from the hydride.

2.9. Simulation variables, parameters, and function used

The mathematical model is applied to the geometry built above and various variables parameter and functions used from the experimental work performed by Jemni et al (1995) are defined specifically and solved with the aid of COMSOL MULTIPHYSICS software. The density of metal hydride presented in the table for this research work is bulk densities (Table 2)

rho_s	6360_[kg/m^3]	Bulk density of metal hydride at saturation
rho_emp	- 6000[kg/m^3]	Bulk density of metal hydride
Cps	- 419[j/kg/k]	Solid- specific heat capacity
Cpg	- 14890[j/kg/k]	Gas- specific heat capacity
k_e	- 1.32[w/m/k]	effective thermal conductivity
K	- 1e_08	permeability
epsilon	0.5	porosity
Mg	2.01588_03_[kg/mol]	hydrogen molecular weight
Rg	- 8.314[J/mol/k]	universal gas constant
ca	- 59.187_[1/s]	absorption constant
cd	- 9.57_[1/s]	desorption constant
Ea	- 21179.6[J/mol]	absorption activation energy
Ed	- 15473_[J/mol]	desorption activation energy
T_ini	- 293_[k]	initial temperature
P_ini	156000_[Pa]	initial pressure
deltaH	-1.539e+07_[J/kg]	heat of formation in abs/des reaction
eta	- 9.2e-06_[Pa*S]	dynamic viscosity
h1	- 150_[w/m^2/k]	heat source
v_out	- 1.098_[m/s]	outlet velocity
Po	- 18_[bar]	initial Pressure
P	- 50_[bar]	Pressure
Pg	- P_[Pa]	gas pressure from Darcy
Cps	- 419_[j/kg/K]	Solid specific heat capacity

Table 2. Variables and function used in simulation (Jemni et al. [3])

The developed mathematical model included the following variables and functions, which were used to simulate heat and mass transfer inside hydrogen storage tank. The variables and functions are listed in the Table 3 below.

T_ini	293_[K]	Initial temperature
p_ini	143,220_[Pa]	Initial pressure
c_ini	rho_emp/Mg	Initial concentration
rho_s	c.*Mg	Current density of solid MH
rho_sat	8527_[kg/m^3]	Saturated density

http://dx.doi.org/10.5772/61625

rho_emp	8400_[kg/m^3]	Empty density
Cps	419_[J/kg/K]	Solid specific heat
Cpg	14890_[J/kg/K]	Gas specific heat
k_e	1.32_[W/m/K]	Effective thermal conductivity
K	1e-08_[m^2]	Permeability
deltaH	-1.539e+07_[J/kg]	Heat of formation in abs/des reaction
epsilon	0.5	Porosity of the gas
Mg	2.01588e-03[kg/mol]	H2 mol weight
Rg	8.314[J/mol/K]	Universal gas constant
Ca	59.187[1/s]	Absorption constant
Cd	9.57[1/s]	Desorption constant
Ea	21179.6[J/mol]	Absorption activation energy
Ed	15473_[J/mol]	Desorption activation energy
Pg	p	Gas pressure
Peqa	1000[Pa]*exp(17.6083704.6[K]/T)	Equilibrium pressure of absorption
Peqd	1000[Pa]*exp(17.4783704.6[K]/T)	Equilibrium pressure of desorption
m_des_fin	if(Pg<Peqd,Cd*exp(Ed/(Rg*T)) *(g-Peqd)/Peqd*(rho_s-rho_emp),0)	Mass source during desorption process
m_abs_fin	if(Pg"/>=Peqa,(Ca*exp(Ea/(Rg*T)) *log(Pg/Peqa)*(rho_sat-rho_s)),0)	Mass source during absorption process
m1	if(abs(m_des_fin)"/ >abs(m_abs_fin),m_des_fin,m_abs_fin)	Auxiliary term
M	if(((m_des_fin==0)&&(m_abs_fin==0)), 0,m1)	Total mass source/sink

Table 3. Simulation variables and functions

2.10. Model verification

To validate the model, the results of Jemni et al.[3] were used. The simulation results obtained were compared with one obtained by Jemni et al. [10]. The comparison of the axial temperature distribution in the paper of Jemni and the calculated results shows agreement with each other. (Figure 4).

Further modification of simulation variables and functions used by Jemni et al. with introduction of some conditions upon simulation produced similar results, which can be shown in the Figures 5 and 6. The main body of results obtained in current research is given in the following Chapter.

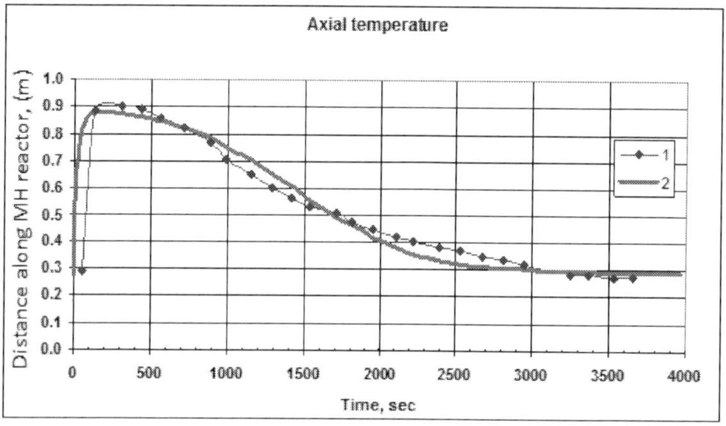

Figure 4. Axial temperature distribution in metal hydride reactor. 1 – Experiment (Jemni et. al. [3]-), 2 – calculations.

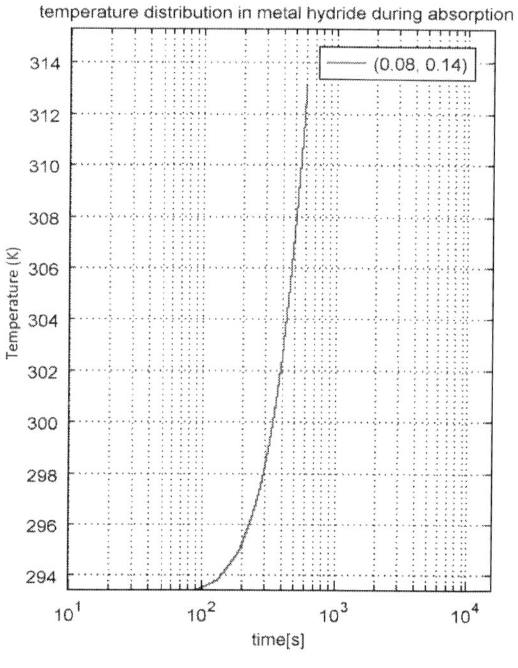

Figure 5. Time evolution of temperature in metal hydride reactor during absorption at point with coordinates [0.08, 0.14] during the 600-s period.

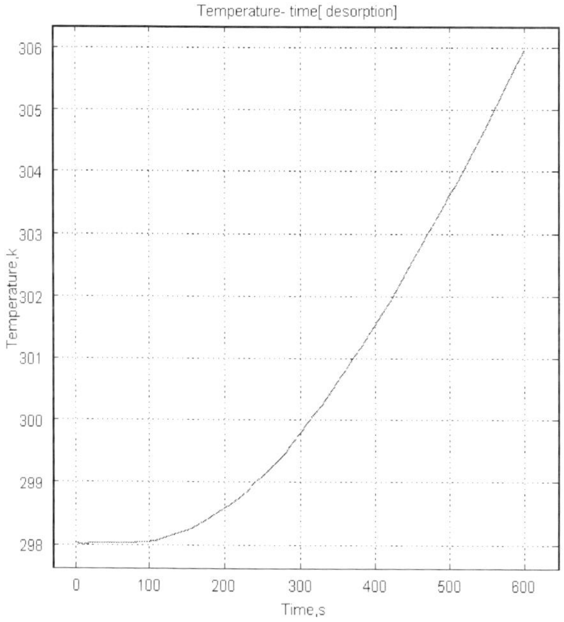

Figure 6. Time evolution of temperature in metal hydride reactor during desorption at point with coordinates [0.08, 0.14] during the 600-s period.

There is a significant - agreement between the results obtained by Jemni et al. [3], and the simulation for this work, for both absorption and desorption process. While Jemni et al. [10] used the cylindrical tank as a reactor for metal hydride hydrogen storage, in this work, most of the simulations were performed for the reactor with - rectangular fins.

3. Results and discussion

3.1. Dynamics of absorption process

Absorption process was simulated for hydrogen storage tank filled with AB_5 type metal hydride. The geometry of the tank and the boundary conditions were described in the sections 3.1 and 3.3. The position of the sensor (measurement point) for temperature, pressure, velocity, and density has the following coordinates: radial coordinate r=0.08 m, axial coordinate z=0.14 m, which is approximately in the center of the metal hydride bed(Domain 1).

3.1.1. Temperature time distribution

Figure 7. illustrates the temperature distribution of hydrogen in MHHST during absorption at point [0.08, 0.14] of the bed. The simulation results show a gradual increase in temperature at different time interval. The total simulation time was taken to be 600_s. The temperature at initial stage of the reaction was very slow but started to increase gradually from 200_s to 600_s, due to the exothermic nature of the reaction between hydrogen and metal hydride. More heat has been release into the bed and this tends to increase the temperature of the heating fluid. A maximum temperature of 313K within the bed was reached.

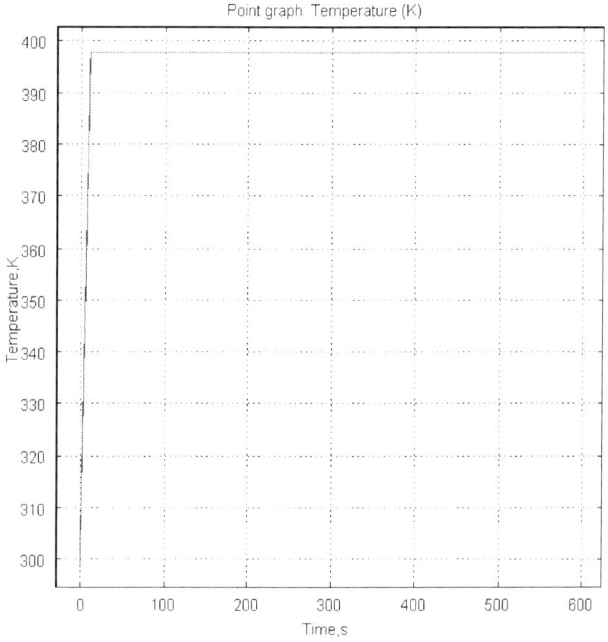

Figure 7. Time evolution of temperature in metal hydride reactor at point with coordinates [0.08, 0.14] during the 600-s period.

3.1.2. *Pressure distribution*

Figure 8. shows the pressure variation with time at point [0.08, 0.14] within the MHHST during absorption reaction of hydrogen. It was observed that there is an increase in the pressure in the bed during absorption process at various simulation times. More hydrogen was absorbed due to increase in temperature and pressure in the MHHST bed, which was due to the exothermic nature of the reaction. The pressure increased gradually from initial stage of the reaction, which was 0_s to final time of 600_s. At this time, the maximum pressure was reached in the bed, which is closer to the equilibrium pressure. The equilibrium pressure is 1000_[Pa]*exp[17.6083704.6[K]/T], in which the temperature T is 273_K, the temperature in the middle of the reactor. Lower temperature yields more hydrogen,and the heat released during reaction slows down subsequent reaction for higher coolant temperature. In addition, the higher the charging pressure, the faster the reaction rate.

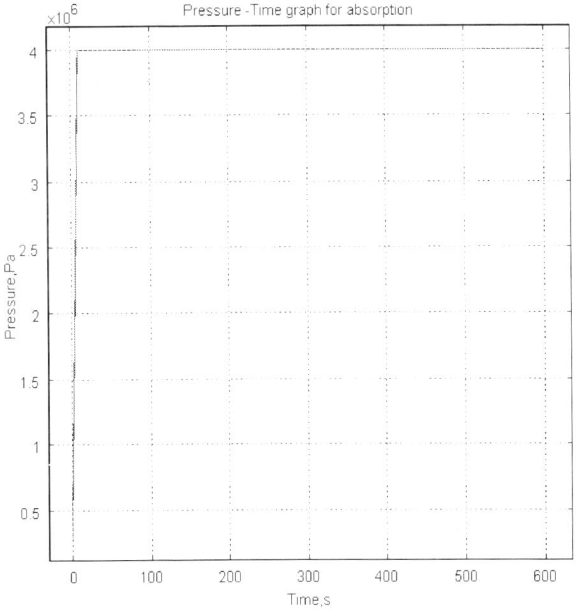

Figure 8. Time evolution of pressure in metal hydride bed at point with coordinates [0.08, 0.14] during the 600-s

3.1.3. Concentration distribution

Figure 9. illustrates the changes in the concentration of metal hydride when it absorbs hydrogen at different time intervals. It was observed that the mass of metal hydride that was loaded inside the bed changes with time gradually. At initial stage of the reaction, the concentration of metal hydride is 4.2, this value reduces to $4.225 x 10^6$ at the final simulation time of 600_s. It can be concluded that at increasing time of the simulation, the concentration of metal hydride in the bed will decrease further when fully absorbed hydrogen.

3.1.4. Hydrogen flow rate during absorption

Figure 10. illustrates the flow rate of hydrogen into MHHST at different time interval during absorption. The flow of hydrogen into the metal hydride tends to increase with time. At the point [0.08, 0.14] of the bed, the flow rate started to increase from 0 to 0.0025kg/s and further increased to the maximum flow at a simulation time of 600s, where there is a larger absorption of hydrogen in MHHST. It can be concluded that flow rate of hydrogen into MHHST tends to increase with an increasing simulation time and a larger heat transfer coefficient values.

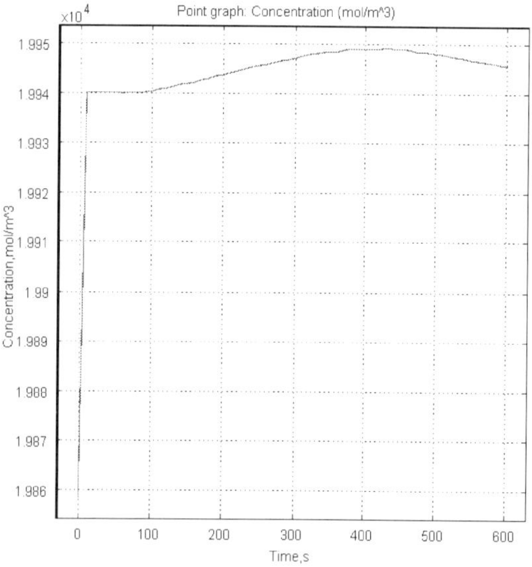

Figure 9. Time evolution of metal hydride concentration at the point with coordinates [0.08, 0.14] during the 600-s period

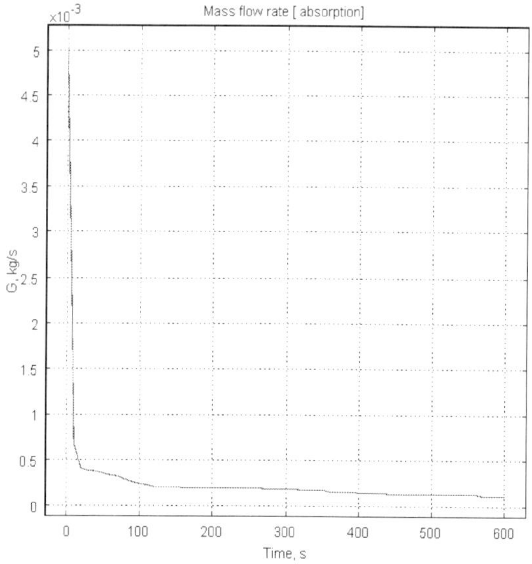

Figure 10. Flow rate of hydrogen in metal hydride during absorption in simulation time 600-s.

3.2. Dynamics of desorption process

3.2.1. Temperature time distribution during desorption

Figure 11 shows, the changes in the temperature at different times during the hydrogen desorption from MHHST. Due to the endothermic nature of the reaction, the desorption of hydrogen at initial stage, which is 0_s is very slow, and the temperature value was noticed to be 293_K, which is the initial temperature of the simulation. As the time increases, the temperature along the MHHST also tends to increase. This is happening up to the 600s. At that time the temperature increases rapidly to much higher values. After a long time the system comes down to thermal equilibrium with the heating fluid. It can be concluded that desorption occurs very fast at higher temperatures with an increase of operating time.

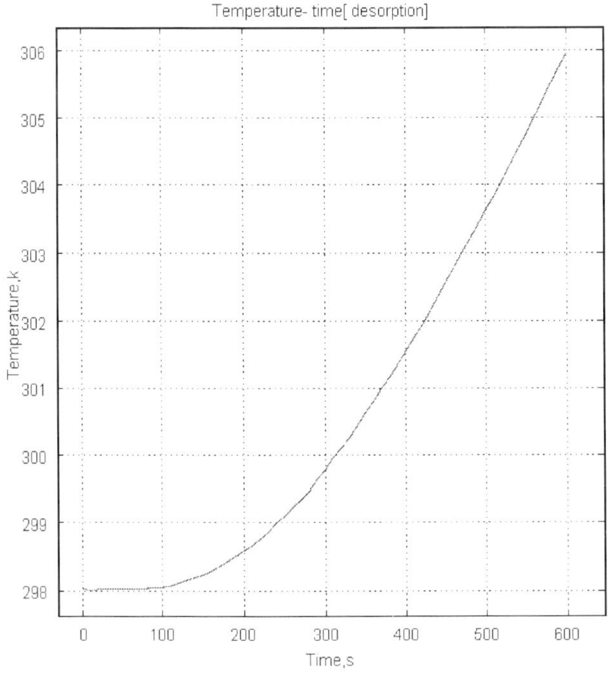

Figure 11. Time evolution of temperature in metal hydride reactor at point with coordinates [0.08, 0.14] during 600s period.

3.2.2. Pressure time distribution during desorption

Figure 12 shows that pressure distributions at the point [0.08, 0.14] within the MHHST during desorption with initial gas phase pressure are set to be 10_bar, which is equal to equilibrium temperature at 293_K (initial bed temperature). Although the MHHST is operated at 8_bar at the start, the simulation shows that before the average pressure approaches the equilibrium value, only the small amount of hydrogen in gas phase is released and there is no desorption in this period. From Figure 12, we can see that there is initial pressure drop from 8bar to the equilibrium pressure. The focus was based on what happened when desorption start (t=0). The discharging pressure is set to be 1_bar.desorption rate increases as the coolant temperature increases, since hydrogen desorption in metal hydride is an endothermic reaction. Moreover the centre temperature drops down right after the desorption reaction occurs.

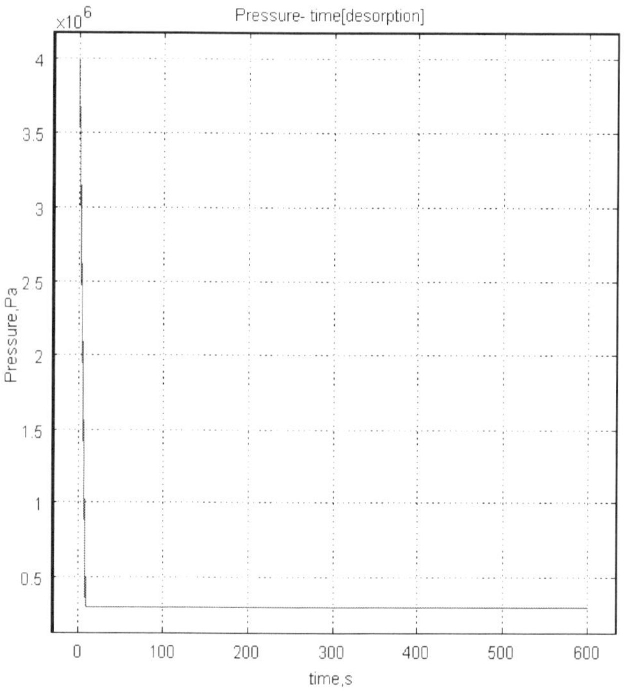

Figure 12. Pressure distribution of hydrogen from metal hydride at [0.08, 0.14] in simulation time 600s during desorption.

3.2.3. *Velocity time distribution during desorption*

In Figure 13, the velocity of hydrogen released during desorption within MHHST is shown at time interval of 600_s. At a point [0.08, 0.14] inside the MHHST, the velocity of hydrogen increases with increasing simulation time due to endothermic nature of the reaction and tends to decrease further as simulation time is increases. If the desorption temperature varies within the bed and pressure is very low, the velocity of hydrogen increases.

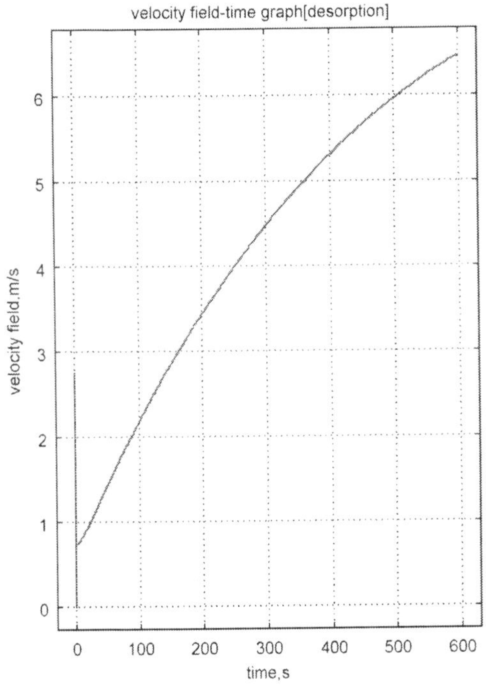

Figure 13. Velocity distribution of hydrogen in metal hydride during desorption.

3.2.4. *Concentration distribution*

The concentration of hydrogen in metal hydride tank during reaction decreases rapidly with time, as shown in Figure 14. There was a larger percentage of hydrogen in metal hydride storage due to the cold nature of the wall and low heat transfer from the bed to the surrounding cooling fluid, and this was due to low poor thermal conductivity of the metal hydride particles. The concentration of hydrogen decreases as the time increases, and heat distribution within the MHHST also increases. At higher temperature and higher heat transfer coefficient, the concentration of hydrogen in metal hydride decreases with increasing simulation time.

3.2.5. *Hydrogen flow rate*

The rate of hydrogen desorption from metal hydride storage tank is shown in Figure 15. At the start of the reaction, hydrogen desorption rate is very slow at 0_ s, but it increases gradually with time as the tank is heated up at a temperature of 333K.more hydrogen was released from metal hydride hydrogen storage tank. At simulation time of 600_s, larger percentage of hydrogen was released from MHHST, and it is evidence that at increasing simulation time and higher temperature, and low pressure, more hydrogen will be released.

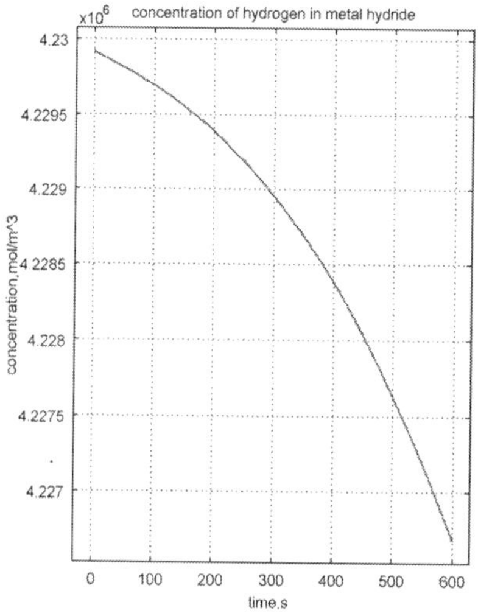

Figure 14. Concentration of hydrogen in metal hydride during desorption at simulation time (600_s).

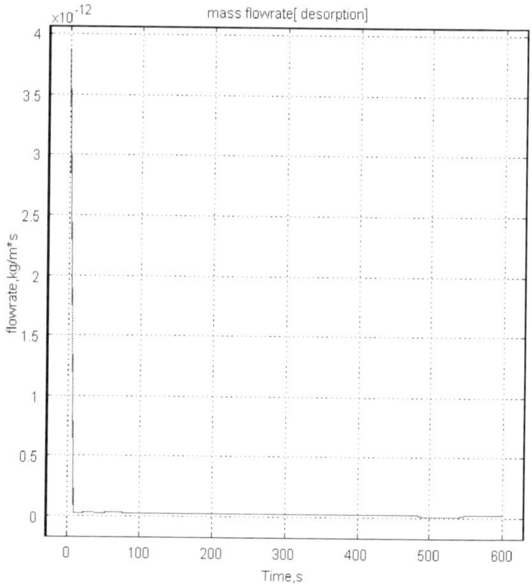

Figure 15. Hydrogen flow rate from metal hydride in simulation time 600s during desorption

http://dx.doi.org/10.5772/61625

3.2.6. *Temperature distribution in the tank with fins*

Figure 16 shows the temperature distribution within the MHHST during simulation of hydrogen absorption and desorption at simulation time of 600_s.the temperature profile shows two different temperature scales supplied to the system. The temperature supplied by the copper plate to the system, which changes with time along the MHHST, the temperature released by metal hydride when it absorbs/ desorbs hydrogen; it was observed that temperature increases rapidly at different distances of the MHHST. The time taken for the temperature to be released was 551.4_s.

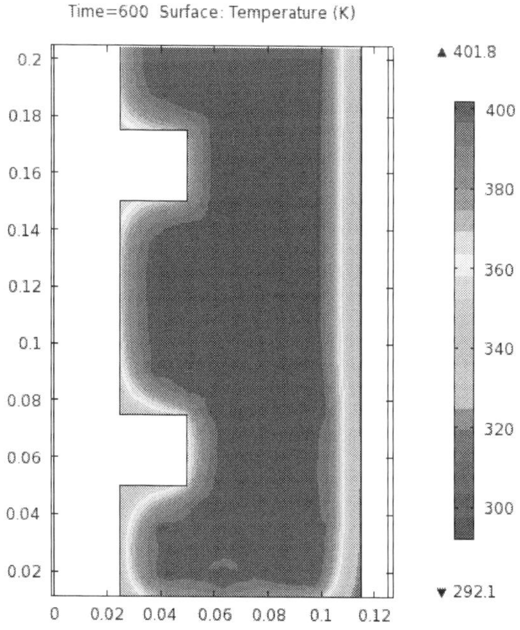

Figure 16. Temperature pattern along the metal hydride hydrogen storage tank.

3.2.7. *Pressure profile*

Figure 17 shows the pressure distribution in MHHST during the simulation, there is an increase in pressure along different distances within the computational domain of the tank. it was observed that pressure increased rapidly from the equilibrium pressure to the set point pressure of 40 bar at a temperature of 273_K within a simulation time of 600_s for absorption process. In addition, for desorption process, the pressure decreases rapidly from the equilibrium pressure to set point pressure of 10bar at a temperature of 333_K within a simulation time of 600s in the tank due to the endothermic nature of the reaction.

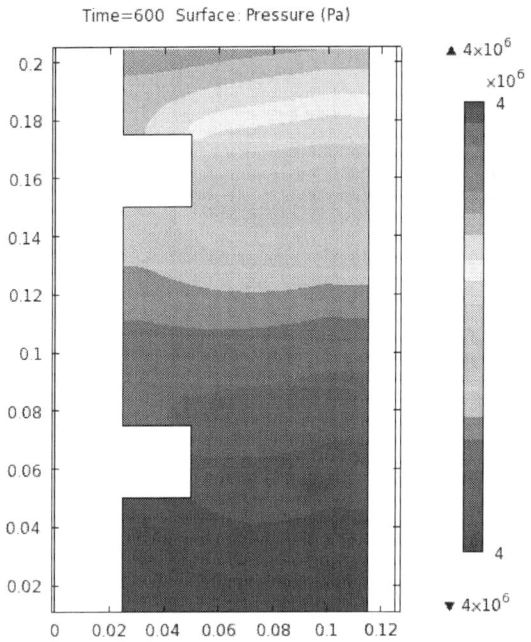

Figure 17. Pressure variation along metal hydride hydrogen storage tank.

4. Conclusions

The objectives of this research have been achieved, thus dynamic mathematical model for the MHHST was developed. The following conclusions can be drawn;

1. In order to verify the model, the numerical model presented above was compared to experimental results of Jemni et al, [3]. The influence of all the operating parameters in the model such as temperature, pressure, density, velocity, and flow rate was investigated. It was found that the pressure and temperature axial distribution in the MHHST for absorption, are in good agreement with the measured experimental values of Jemni et al. [3].

2. The simulations of heat and mass transfer for MHHST with different internal geometries were done. Heat transfer, mass transfer and momentum transfer modules were effectively incorporated in COMSOL Multiphysics 4.0a software. Geometry was built for the problem, and boundary conditions were properly allocated to the each domain in the model. The energy (convection and conduction), mass (diffusion), and momentum (Darcy's law) transport equations and some specific user – defined subroutines were

implemented in the software. It was discovered that the geometry with fins results in faster absorption and desorption when compared with the geometry without fins, in which the absorption/desorption rate is very slow and produce very negligible values for absorption/desorption rate.

3. Temperature, pressure and density non-steady distributions inside the metal hydride porous bed were revealed. It was found that absorption occurs faster at a lower temperature and higher pressure because absorption process is exothermic reaction, and desorption occurs at higher temperature and lower pressure due to the endothermic nature of desorption process. Concentration increases with absorption of hydrogen and decreases with desorption of hydrogen.

5. Abbreviations

2D; Two-Dimensional

PDE; Partial differential Equations

MHHST; Metal hydride hydrogen storage tank

USDE; United States Department of Energy

CFD; Computational Fluid Dynamic

DOE; Department of Energy

PEMFC; Proton exchange membrane fuel cell

ODEs; Ordinary differential equations

HHV; Higher heating value

LHV; Lower heating value

SPEFC; Solid polymer electrolyte fuel cell

MFPM Multi; facilitated proton membrane

DFAFC; Direct formic acid fuel cell

Acknowledgements

The authors will like to express their sincere gratitude to the faculty of Engineering and built environment, Tshwane University of Technology Pretoria South Africa for her financial assistance.

Author details

Olaitan Akanji* and Andrei Kolesnikov

*Address all correspondence to: laitan436us@yahoo.com

Department of Chemical, Metallurgical & Materials Engineering, Tshwane University of Technology, Pretoria, South Africa

References

[1] Visaria, M., Mudawar,l.,Poupoint,T.,Kumar,S.Int.J.HeatMass Trans.2010,53, 2229 -2239.

[2] Zhixiong,G., Hyung,J.S, J.lnt.j.. Heat and Mass Trans.1999,42,379-382.,

[3] JEMNI, A., BEN NASRALLAH, S. Study of two dimensional heat and mass transfer-During adsorption in a metal- hydrogen reactor, *Int. J. Hydrogen Energy 1995,* 20, 43-52.Doi:10.1016/0360-3199(93) E0007-8.

[4] Aldas, K., MAT M.D. AND KAPLAN, Y. A three dimensional mathematical model for absorption in a metal hydride bed, *Int. J. of Hydrogen Energy 2002,* 27 1049-1056.Doi/10.1016/j..ijhydene.2011.12.140

[5] Mayer, U., Grill, M. AND Supper, W. Heat and mass transfer in metal hydrides reactions beds, experiment and theoretical results, *J. Less-common Metals 1987,* 131, 238.

[6] Nakagawa, T., Inomata, A., Aoki, H. & Miura, T. Numerical analysis of heat and mass transfer characteristics in the metal hydride bed. Int. j. Hydrogen Energy 2002, 25,4,339-350.Doi:-org/10.1016/so360-3199 (99) 0036-1

[7] Dhaoua, H., Mlloulia, S., Askaria, F., Jemni, A. Ben Nasrallahi, S. Int.J.Hydrogen Energy 2007, 32, 1922.Doi:/10.1016/j.ijhydene.2006.08.045.

[8] Jemni, A., Nasrallah, SB, Lamloumi, J. Experimental and theoretical study of a metal hydrogen reactor, int. j. of Hydrogen Energy, 24, 631- 644.

[9] Chung, C.A.., HO,C. J Thermal fluid behavior of the hydriding and dehydriding processes in a metal hydride hydrogen storage. Doi-10.1016/SO360-3199 (01) 0030-1.

[10] MAT, M.D. AND KAPLAN, Y. Numerical study of hydrogen in a La-Ni5 hydride Reactor, *Int.J.Hydrogen Energy2001,* 26, 957.

Section 4

Hydrogen Utilization

Chapter 5

Treatment of Hydrogen Bonds in Protein Simulations

Ya Gao, Ye Mei and John Z. H. Zhang

Additional information is available at the end of the chapter

http://dx.doi.org/10.5772/61049

Abstract

The hydrogen bond plays an essential role in maintaining the secondary structures of protein, and an accurate description of hydrogen bond interaction is critical in protein folding simulation. Modern classical force fields treat hydrogen bonding as nonbonded interaction, which is dominated by electrostatic interaction. However, in the widely used nonpolarizable force fields, atomic charges are fixed and are determined in a mean-field fashion. Applying nonpolarizable AMBER force field in the folding simulations of some short peptides, the native structure cannot be well populated. When polarization effect is introduced into the simulation by utilizing either the on-the-fly charge fitting or the polarizable hydrogen bond model, the native structure becomes prominent in the free energy landscape. These studies highlight the necessity of electrostatic polarization effect in protein simulation.

Keywords: hydrogen bond, quantum mechanics, molecular mechanics

1. Introduction

The function and chemical properties of protein are determined by its three-dimensional structure.[1, 3] The native structure of protein at physiological temperature and solvent condition, which is located at a free energy minimum, is a result of delicate balance between many competing interactions, including van der Waals forces, hydrophobic interaction, and electrostatic interaction, such as salt bridge and hydrogen bond (H-bond).[4, 5] During folding process, the unfolded structure collapses first to minimize the area of hydrophobic surface that is exposed to the solvent and hydrogen bonds begin to form. Hydrogen bonds, providing most of the directional interactions, are the dominant feature in the formation of protein secondary or tertiary structures. They play an indispensable role in the stabilization of the native structure of a protein. α helix and pleated β -sheet were predicted in 1951 by Linus Pauling and Robert Corey on the basis of hydrogen bonding and cooperative criteria.[6] Therefore, correctly describing the thermodynamic properties of inter/intra-protein hydrogen bonds is essential.

Hydrogen bond is formed when an electronegative atom (H-bond acceptor, A) approaches a hydrogen atom bonded to another electronegative atom (H-bond donor, D). There exist many hydrogen bond donors and acceptors in proteins, such as the amide and carbonyl groups of the backbone, as well as the polar functional groups (amides, carboxyl groups, hydroxyl groups, and amines) in the side chains of some amino acids.

Different parameters can shed light on the different aspects of the nature of H-bonds established for a given system. The interaction energy is the best indicator of H-bond strength. The geometry of the H-bond complex, and in particular, the distance and angle around the H-bond also provide a very good indication of such interaction. When a hydrogen bond is formed, the distance between the H and the acceptor (A) atoms has to be smaller than the sum of their corresponding van der Waals radii. Furthermore, the D–H–A angle should be close to 180º. In practice, a cutoff of 120 or 90º is employed. Now, it is well accepted that, in neutral systems, H-bonds formed between O, N, or O/N pairs of atoms are very strong, whereas those formed between C-H and a π system are weak. The formation of an H-bond can usually cause red shift of IR spectroscopic bands, even though improper blue shifts have been also detected.

2. Computational studies of hydrogen bonds

2.1. Quantum chemistry method

Quantum chemistry method is an ideal approach to calculate the energy of a hydrogen bonded system since all electronic and steric effects are fully taken into account. Molecular orbital methods, especially the post Hartree–Fock (HF) methods, are capable of providing an accurate description of systems containing hydrogen bonds. Thus, Møller–Plesset methods such as MP2[7] can be considered ideal for the study of hydrogen bonds. However, even a single point calculation at MP2 level of medium-sized hydrogen bond systems can be very demanding. Density functional theory (DFT) methods such as B3LYP[8, 9] is becoming more and more popular these days. It has been shown that B3LYP calculation provides very good description of strong H-bond systems, comparing with the experimental data or MP2 results. However, Kim et al. have shown that in the case of some systems with weak H-bond such as C-H π or O-H π, B3LYP does not offer the same level of reliability as MP2.[10]

The basis set superposition error (BSSE) is another issue to be considered in the computation of H-bond complexes, which arises because of the incompleteness of the basis set functions. Applying the basis set of the complex to each molecule will yield a lower energy than that from the calculation with the basis set for the monomer itself. Therefore, BSSE introduces a non-physical attraction between the two monomers. It has been shown that the BSSE correction of the potential energy surface leads to sizeable differences in the H-bond lengths for correlated calculations using split-valence basis sets with polarization and diffuse functions.

Although much progress has been made for investigating the nature of H-bonds utilizing quantum chemistry methods, few *ab initio* calculations have been carried out on the calculation of dynamic properties of hydrogen bond interaction for biomolecules due to relatively large size of systems. Thus, a more efficient computational method is desired.

2.2. Molecular mechanics method

Molecular mechanics is a popular computational method, in which the electronic degrees of freedom of the molecules are ignored, and only motions of the nuclei are calculated. The fundamental assumption underlying all molecular mechanics methodologies is the Born–Oppenheimer approximation, which allows separated motions for nuclear and electrons.

In molecular mechanics methods, the molecule of interest and surrounding solvent are typically treated as classical particles interacting through an empirically derived and computationally tractable potential energy function ("force field"). Molecular dynamics or Monte Carlo simulations can provide wealth of information to interpret, complement and design experiments. Most force fields were developed in the early 1980s, such as AMBER, CHARMM, and OPLS. In the first-generation AMBER force field, hydrogen bond energy is explicitly expressed as a 10–12 function as $\frac{C_{ij}}{R_{ij}^{12}} - \frac{D_{ij}}{R_{ij}^{10}}$. The inclusion of the first term is to prevent the occurrence of unrealistically short H-bonds and the second term is to "fine tune" the H-bond distances and energies to desired values.

This kind of potential function had not been changed until AMBER94 force field was proposed. In AMBER94, hydrogen bond is incorporated into electrostatics interaction due to the improved performance of the new RESP charge model and new van der Waals parameters. Current AMBER force fields, following the potential energy functional form of AMBER94, comprise bonded (bond, angle, and dihedral) and nonbonded terms (van der Waals and electrostatic interaction), which are parameterized against gas phase quantum chemical calculations, spectroscopic, and thermodynamics experimental data of small model molecules. The potential energy function has the following form.

$$
\begin{aligned}
V(r) &= \sum_{bonds} K_b \left(b - b_0\right)^2 + \sum_{angles} K_\theta \left(\theta - \theta_0\right)^2 \\
&+ \sum_{nonbij} \left(A_{ij} / r_{ij}^{12}\right) - \left(B_{ij} / r_{ij}^{6}\right) + \left(q_i q_j / r_{ij}\right) \\
&+ \sum_{dihedrals} \left(V_n / 2\right)\left(1 + \cos\left[n\phi - \delta\right]\right)
\end{aligned}
$$

3. Treatment of hydrogen bond in protein simulations

3.1. Deficiency in the treatment of hydrogen bond

Utilizing molecular dynamics simulations to investigate the structural properties of proteins has been an important tool in biological science for decades. With continuous advancing of computer technology, the difficulty in time scale for protein simulation is lessened significantly. For example, D. E. Shaw utilized a specialized supercomputer, Anton, to investigate the folding mechanism of a series of proteins, and the simulation time was extended to microseconds.[11] A 58-residue protein BPTI conducted by Shaw also reached one millisecond.[12] However, successful application of computer simulation is impeded by the accuracy of force field employed. Defects in the potential energy could strongly bias the simulation results toward incorrect conformations.

An important deficiency that exists in current pairwise force fields is lack of polarization.[13, 16] When an isolated molecule interacts with another molecule, or is placed in an external electric field, its charge distribution will be distorted due to polarization effect. Current force fields are amino-acid specific or mean-field like and, therefore, fail to give accurate representation of the electrostatics of the specific protein environment, which is highly inhomogeneous and protein specific. In a previous work of Zhang et al.,[17] a short α -helix cannot fold to its native structure with AMBER03 [18] force field, despite that this force field has been proved to bias helical conformation.[19] In the past two decades, many attempts have been made to explicitly incorporate polarization effects into molecular modeling,[20] such as the fluctuating charge model,[21, 22] Drude oscillator,[23, 24] induced multipole,[25, 29] and quantum mechanical treatment.[30, 33]

3.2. Improvement in the treatment of hydrogen bonds

Based on a recently proposed fragment-based scheme (molecular fractionation with conjugate caps, MFCC), a new scheme termed polarized protein-specific charge (PPC) model has been proposed for protein dynamics.[34] PPC is derived from first principle quantum calculation and therefore can correctly represent the electronically polarized state of the protein. It has been proven that the polarization effect plays an important role in pK_a shifts for ionizable residues,[34] hydrogen bond stability,[35, 40] protein folding, and native structure stabilization,[17, 41, 45] dynamic properties of proteins,[46, 48] protein–ligand binding affinity,[47, 49, 51] and protein–protein recognition specificity.[52] The polarization of backbone hydrogen bonding is found critical to the success of folding simulation of proteins. In those works, the polarization of hydrogen bonds is included in the simulations by performing on-the-fly quantum mechanical calculations to generate dynamically updated atomic charges. However, on-the-fly charge fitting is computationally expensive, in which quantum mechanical calculations must be periodically carried out to obtain new atomic charges for residues involved in hydrogen bond formation or cleavage. It is desirable to employ a simpler but effective method to include polarization of hydrogen bond for protein simulation. Thus, a new polarization model for the main chain hydrogen bond based on an analytical fitting of atomic charges was proposed, which is computationally more efficient.[53]

In the fitting, a pair of alanine dipeptides with a backbone hydrogen bond was utilized as the model system, which is shown in Figure 1.

Atomic charges are calculated by a restrained fitting to the electrostatic potential generated from quantum chemistry calculation.[54, 56] By varying the donor–acceptor distance but keeping the other degree of freedom fixed, the charge variation of atom N and O with respect

http://dx.doi.org/10.5772/61049

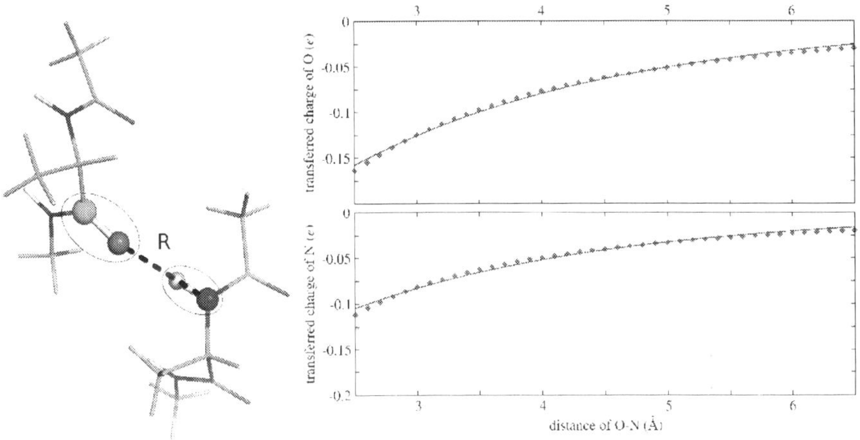

Figure 1. Left: model system used in fitting the atomic charge polarization as function of donor–acceptor distance in hydrogen bond pair. The charge transfer is only allowed between N and H, and between C and O as dipeptide in the ellipses. Right: amount of charge transfer as a function of donor–acceptor distance of atoms O (top) and N (bottom).

to their asymptotic values are shown in Figure 1, both of which can be fitted to a single exponential function. In this approach, charge transfer between residues is not allowed and only the atomic charges of the N, H, C, and O atoms in the hydrogen bond can vary. Thus, the amounts of charge transferred for H and C atoms are exactly the negative of that for N and O atoms, respectively. The fitted functional forms for N and O atoms are

$$\Delta q_N = -0.493 \times \exp(-0.455R),$$

and

$$\Delta q_O - 0.743 \times \exp(-0.466R),$$

in which R is the donor–acceptor distance in angstroms. As shown in Figure 1, at the normal hydrogen bond length (~3 Å), the transferred charge is only about 0.1 e and 0.13 e for N and O atoms, respectively. In the charge fitting procedure, mutual polarization was included. Specifically, in the calculation of one dipeptide, the other one was taken as background charges centered at atomic positions. The calculation was iterated until convergence was reached. When the two dipeptides are infinitely separated, the residues are given the corresponding AMBER charges. Thus, the charge variation can be determined by hydrogen bond length alone. Compared with PPC model, this polarized hydrogen bond (PHB) charge model is more efficient. In the current approach, only the main chain hydrogen bonds are considered. Hydrogen bonds involving side chain atoms, as well as salt bridges, also have strong electrostatic polarization effect, which may be taken into consideration by refined versions of this approach.

4. Applications

4.1. Folding of a short helix under the PPC model

Electrostatic polarization and charge transfer play important roles in protein folding.[57, 58] However, the fixed PPC from a prefixed structure cannot present the dynamic properties of protein accurately, which undergoes a large conformational change in folding. A straightforward method is performing on-the-fly quantum calculation to generate dynamically updated atomic charge at each step of the molecular dynamics simulations. However, the computational demanding is too expensive. A good approximation is implementing polarization into molecular simulations when hydrogen bonds form or break, which is a good indicator of secondary structure. During simulations, only residues involved in the formation or breaking of hydrogen bonds are subjected to quantum calculation to refit atomic charges, while charges of the rest residues are kept to their previous values.

The folding of a short α -helix 2I9M,[59] which is a "de novo"-designed 17-residue peptide, was investigated by employing the dynamic PPC model in molecular dynamics simulations. For comparison, REMD simulation under AMBER03 force field is also performed. Starting from a linear structure, the peptide was successfully folded to its native state with the low backbone root mean square deviation (RMSD) under PPC model. In contrast, due to the lack of polarization effect, the hydrogen bonds of the peptide are too "loose" to form stable secondary structure under AMBER03 force field, as shown in Figure 2.

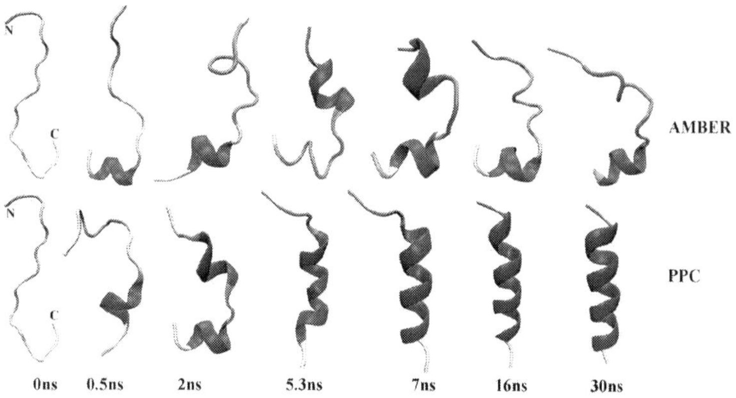

Figure 2. Snapshots of the intermediate structures of peptide 2I9M in simulations using AMBER (upper) and dynamically polarized charge (lower). α-helix, purple; coil, white; turn, cyan. Reprinted with permission from ref. 17. Copyright 2010, American Chemical Society.

Thus, polarized protein-specific charge, especially polarized hydrogen bond charge, plays a critical role in the formation of secondary structures.

4.2. The thermal stability of a short helix under the PHB model

REMD simulations with 12 replicas (temperatures 267, 283, 300, 328, 353, 380, 409, 439, 471, 506, 542, and 577 K) were also performed for investigating the impact of H-bonds on the thermal stability of the peptide. Comparing with standard molecular dynamics simulation at room temperature, the energy barrier can be surmounted much more easily at high-temperature replicas.[60] Two sets of MD simulations were performed using, respectively, the standard AMBER03 and the PHB model. In the PHB model, only atomic charges of donors and acceptors in hydrogen bonds are different, whereas other force field parameters were taken from the AMBER03 force field. The starting conformation was a linear structure generated using LEaP module in AmberTools and relaxed using energy minimization. The relaxed structure was heated to its target temperature for each replica in 100 ps. The solvation effect was modeled using the generalized Born (GB) model developed by Onufriev et al.[61] MD simulations were extended to 120 ns for each replica.

The structure with RMSD from NMR structure below 2 Å was classified as the folded state. The populations of the folded structure in two REMD simulations show a clear difference, as shown in Figure 3.

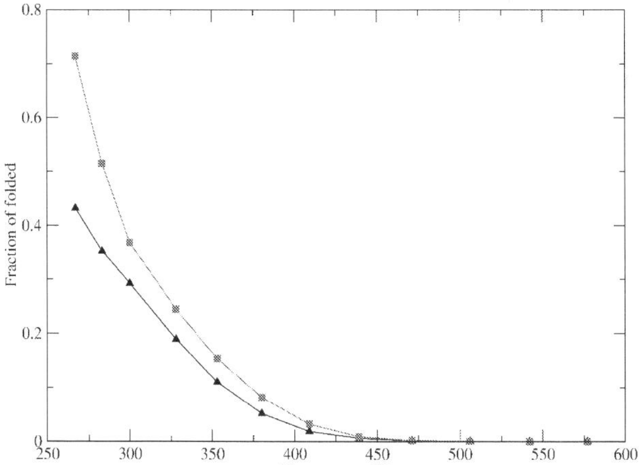

Figure 3. Fraction of folded structure defined by root-mean-square deviation from the NMR structure below 2 as a function of temperature in the simulations utilizing AMBER03 charge (black) and PHB charge (red). Reprinted with permission from ref. 53. Copyright 2012, American Chemical Society.

At the lowest temperature (267 K), the percentage of folded structure using PHB charge model was nearly 30% higher than that in AMBER03 force field. At NMR experimental temperature

(283 K), the fraction of folded structure in the PHB model is over 50%, whereas that in AMBER03 force field is only about 35%, despite the overestimated helical propensity of AMBER03 force field.

The melting temperature is defined as the temperature at which the free energies of the native and unfolded states are equal, indicating the equal population (50% each) of folded and unfolded states. In Figure 3, the melting temperature under AMBER03 force field is well below the NMR temperature of 283 K, which is obviously inconsistent with the experimental observation, suggesting 2I9M is thermally unstable. While under PHB model, the melting temperature is about 286 K, which is in agreement with the experiment. This result shows that hydrogen bond polarization is critical to the thermal stability of the helical structure of 2I9M.

The distribution of the peptide structure from the REMD simulation is determined by the free energy distribution, which is calculated by Weighted Histogram Analysis Method (WHAM) [62, 63] using density of states. Figure 4 plots the free energy changes along the RMSD at 267, 283, and 300 K using both the AMBER03 and PHB model.

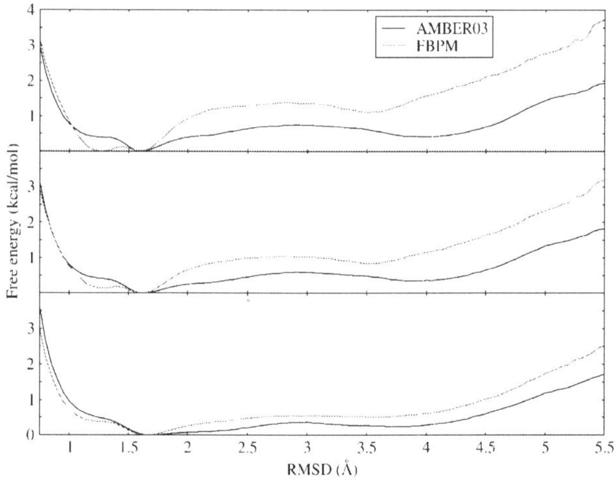

Figure 4. Free energy curves as a function of the RMSD from the NMR structure in simulation utilizing AMBER03 charge (black) and PHB charge (red) at (top) 267 K, (middle) 283 K, and (bottom) 300 K. Reprinted with permission from ref. 53. Copyright 2012, American Chemical Society.

The free energy curve under AMBER03 force field is first examined. At 267 K (top figure), the free energy curve has a global minimum at RMSD around 1.6 Å and also a local minimum at an RMSD value around 4 Å (unfolded state). The energy difference between the global minimum and the local minimum is only 0.5 kcal/mol, indicating comparable populations of structures are distributed at these two geometries. While the free energy curve under the PHB model displays two nearly degenerate minima at RMSD near 1.25 and 1.6 Å, both representing the folded structures. The free energy is essentially flat between the two minima. While at

RMSD = 3.5 Å, the free energy curve shows a very shallow local minimum, which is about 1 kcal/mol higher than the global minimum. This free energy curve explains why the helix is thermally more stable under the PHB model. By increasing temperatures, the free energy curve becomes flatter and the minimum at 1.25 Å gradually disappears but the minimum at 1.6 Å remains, as shown in the middle and bottom figures of Figure 4. While under AMBER03 force field, comparing the free energy curves at 283 and 300 K with those at 267 K, we observe that the free energy difference between the folded and the unfolded states gradually diminished.

The above analysis implies the importance of polarization of backbone hydrogen bonding to the helix formation and the thermal stability of the helix, which is consistent with the observed effect of hydrogen bond cooperativity.[64, 67] The distributions of the number of hydrogen bonds at 267, 283, and 300 K in both REMD simulations are plotted in Figure 5. The number of hydrogen bond is counted using a sigmoidal function of the donor–acceptor distance *d* as related to the strength of hydrogen bonds. It shows that the peak shifts from 3.8 under the AMBER03 charge to 5.0 under the PHB model at 283 K, indicating that hydrogen bonds are strengthened by including the polarization effect. Decreasing the temperature to 267 K, the difference between the AMBER03 charge and PHB model becomes more remarkable, whereas increasing the temperature to 300 K does the opposite.

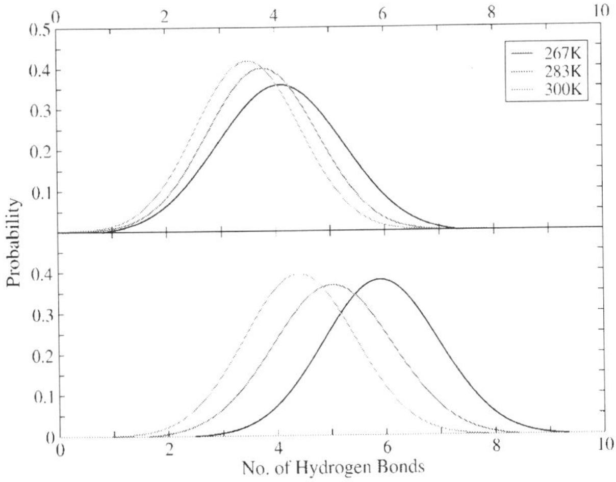

Figure 5. Distribution of native hydrogen bonds population in 267, 283, and 300 K trajectories utilizing the AMBER03 charge (top) and PHB model (bottom). Reprinted with permission from ref. 53. Copyright 2012, American Chemical Society.

4.3. Folding of a long helix under the PHB model

2KHK is a 53 amino acid straight long helix belonging to b30 to 82 domain of F1Fo adenosine-5-triphosphate synthesis in *Escherichia coli*. In order to minimize the overall size of the water box in MD simulations with explicit water, a 10-ns MD simulation with implicit water was performed first. The last snapshot from this short MD simulation was extracted as the initial structure (shown in Figure 6) for explicit water simulation. Simulations were performed using both the standard AMBER03 force field and the PHB model, respectively.

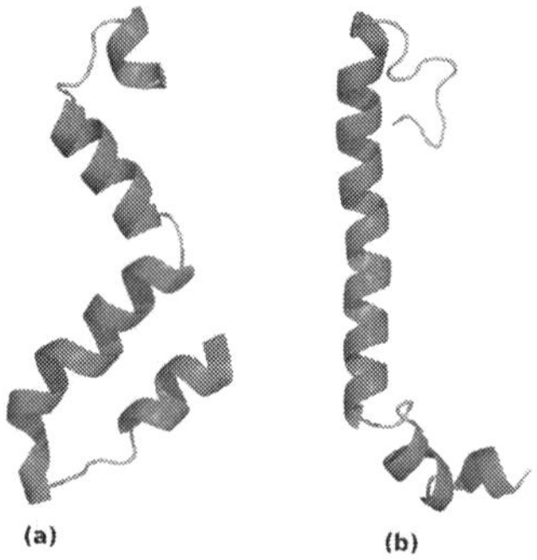

Figure 6. The last snapshots of implicit water simulations (a) under AMBER03 force field and (b) PHB model. Reprinted with permission from ref. 45. Copyright 2013, AIP Publishing LLC.

The variations of RMSD from the native structure and radius of gyration (Rg) along the simulation time are shown in Figure 7.

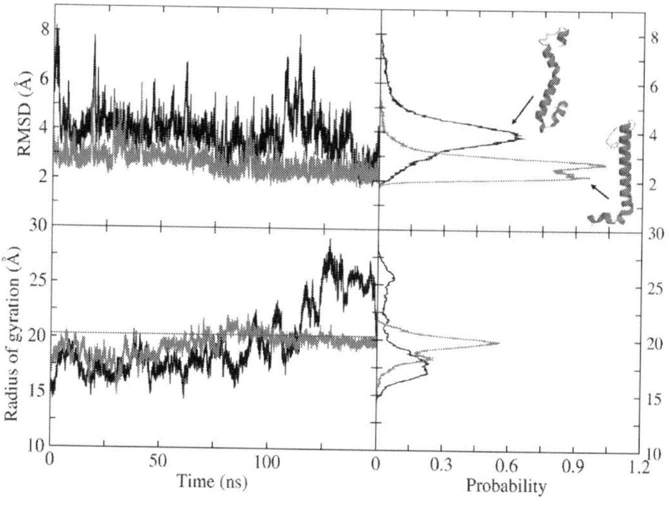

Figure 7. Backbone RMSD from the native structure and radius of gyration as a function of simulation time using AMBER03 charge (black) and the PHB (red) model in explicit water. The representative structures and their relative distributions from these two simulations are also plotted (right panels). Reprinted with permission from ref. 45. Copyright 2013, AIP Publishing LLC.

http://dx.doi.org/10.5772/61049

The overall fluctuations of RMSD and Rg using AMBER03 charges are both larger than those using PHB model. The most populated state under PHB model has an RMSD of 2.1 Å, while under standard AMBER simulation, it is as large as 4.0 Å. The result indicates that hydrogen bond polarization helps the folding of long helix, which is also reflected from the fluctuation of Rg along the simulation time. Under PHB simulation, initially the peptide adopts a more compact structure than the native structure. After 75 ns, the Rg wanders around the experimental value with small fluctuation. While under AMBER03 force field, Rg gradually deviated from the native state. Snapshots of the intermediate states along the simulation time under PHB model are plotted in Figure 8.

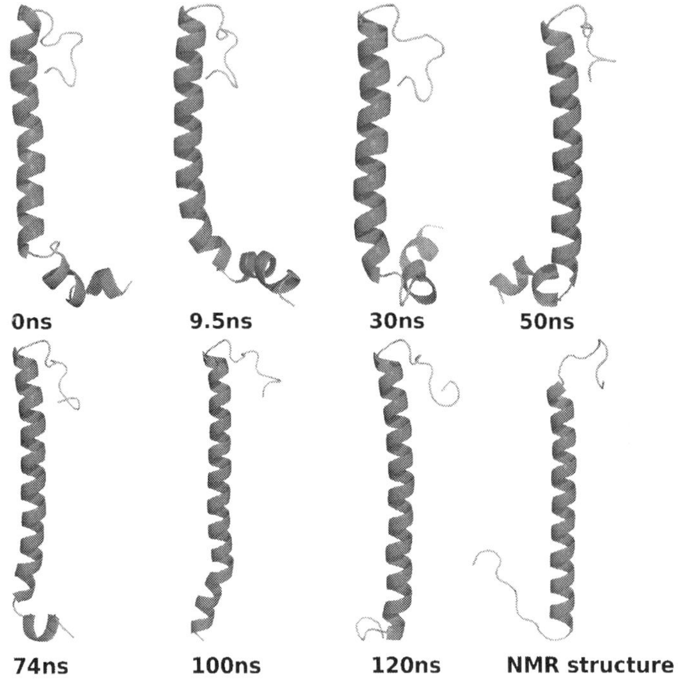

Figure 8. Snapshots of intermediate structures of the peptide at different simulation time under PHB model (compared to the NMR structure). The C-terminal is always on the top. Reprinted with permission from ref. 45. Copyright 2013, AIP Publishing LLC.

It is noted that although the structure near the N-terminal of the peptide is flexible, the helix in the middle region is becoming more and more stable as simulation goes on.

The two-dimensional free energy landscapes from MD simulations under both AMBER03 force field and PHB model are shown in Figure 9, using the RMSD and Rg of the backbone as the reaction coordinates.

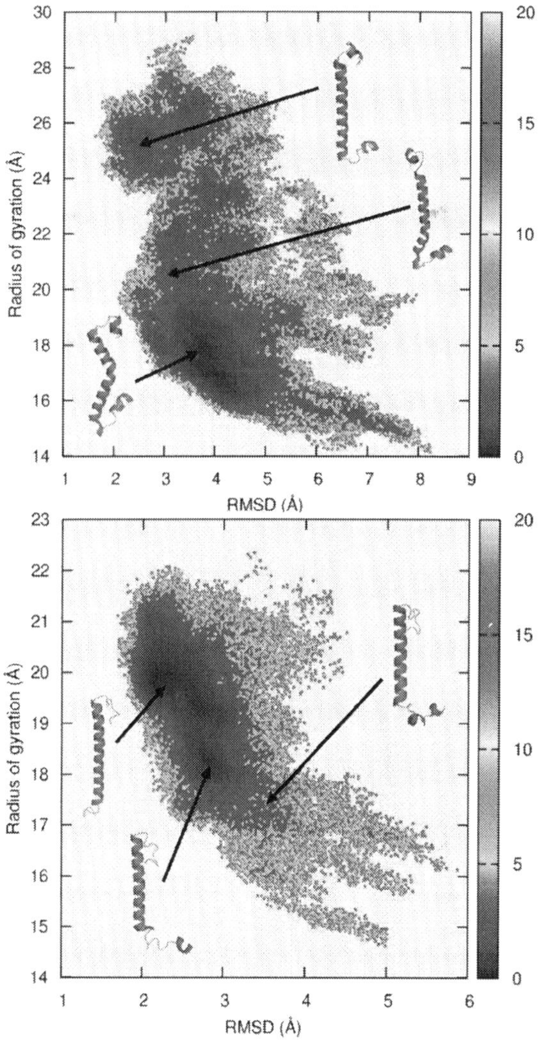

Figure 9. Free energy landscapes at 288 K mapped to RMSD from the native structure and radius of gyration obtained from simulations employing (top) AMBER 03 charge and (bottom) PHB model. The representative structures are also shown in the figure. Reprinted with permission from ref. 45. Copyright 2013, AIP Publishing LLC.

Under AMBER03 force fields, due to the lack of polarization effect, enthalpy change may not be enough to compensate for the entropy loss in the formation of secondary structure. The global minimum located at RMSD of around 4.1 Å and Rg around 17.2 Å (versus 20.3 Å calculated from the nuclear magnetic resonance (NMR) structure). The representative structure is obviously twisted from the middle of the peptide as shown in Figure 9 (top panel). Therefore, no stable native structure is founded in standard AMBER simulation. For comparison, the free energy landscape under the PHB model displays a global minimum at RMSD around 2.0 Å (folded state), as shown in Figure 9 (bottom panel). The global minimum obtained in the PHB model has an Rg value of about 20.0 Å, which is very close to the NMR value of 20.3 Å. There is also a local minimum at RMSD around 3.1 Å, which is a partially unfolded state. At this local minimum, the N terminal of the peptide is unfolded, but the middle helical fragment is almost intact. This free energy analysis shows that the helix is thermally more stable under the PHB model.

5. Summary

Among many factors that are important to protein stability, hydrogen bonds are found to correlate highly with the thermal stability. They are the dominant feature in protein secondary structures of α helices and β sheets. Recent theoretical studies demonstrated that the strength of hydrogen bonds from simulations under standard nonpolarizable force fields is underestimated due to the lack of polarization effect.

In this chapter, we present an accurate and efficient polarizable hydrogen bond model that can be implemented in molecular dynamics simulations. By investigating the thermal stability of a short α helical peptide and the folding of a long α helical peptide, it shows that including polarization effect is very important in maintaining the hydrogen bonds, stabilizing the structure, and folding protein to its native conformation. It also should be noted that folding simulations of peptides with β sheets or mixed helix-sheet structures are still very challenging. The difficulty lies in not only the lack of polarization effect but also the unbalanced secondary structure propensity in used force fields.

Author details

Ya Gao[1], Ye Mei[2,3*] and John Z. H. Zhang[2,3]

*Address all correspondence to: ymei@phy.ecnu.edu.cn

1 College of Fundamental Studies, Shanghai University of Engineering Science, Shanghai, China

2 Center for Laser and Computational Biophysics, State Key Laboratory of Precision Spectroscopy, Department of Physics and Institute of Theoretical and Computational Science, East China Normal University, Shanghai, China

3 NYU-ECNU Center for Computational Chemistry at NYU Shanghai, Shanghai, China

References

[1] Anfinsen, C. B. *Biochem. J.* 1972, *128*, 737.

[2] Pauling, L.; Corey, R. B.; Branson, H. R. *Proc. Natl. Acad. Sci. U. S. A.* 1951, *37*, 205.

[3] Zhang, Y. *Curr. Opin. Struct. Biol.* 2008, *18*, 342.

[4] Chen, J.; Im, W.; Brooks, C. L. *J. Am. Chem. Soc.* 2006, *128*, 3728.

[5] Hubbard, R. E.; Kamran Haider, M. In *eLS*; John Wiley & Sons, Ltd: 2001.

[6] Pauling, L.; Corey, R. B.; Branson, H. R. *Proc. Natl. Acad. Sci.* 1951, *37*, 205.

[7] Møller, C.; Plesset, M. S. *Phys. Rev.* 1934, *46*, 618.

[8] Lee, C.; Yang, W.; Parr, R. G. *Phys. Rev. B* 1988, *37*, 785.

[9] Becke, A. D. *J. Chem. Phys.* 1993, *98*, 5648.

[10] Kim, K. S.; Tarakeshwar, P.; Lee, J. Y. *Chem. Rev.* 2000, *100*, 4145.

[11] Lindorff-Larsen, K.; Piana, S.; Dror, R. O.; Shaw, D. E. *Science* 2011, *334*, 517.

[12] Shaw, D. E.; Maragakis, P.; Lindorff-Larsen, K.; Piana, S.; Dror, R. O.; Eastwood, M. P.; Bank, J. A.; Jumper, J. M.; Salmon, J. K.; Shan, Y.; Wriggers, W. *Science* 2010, *330*, 341.

[13] Guallar, V.; Jarzecki, A. A.; Friesner, R. A.; Spiro, T. G. *J. Am. Chem. Soc.* 2006, *128*, 5427.

[14] Ren, P.; Ponder, J. W. *J. Comput. Chem.* 2002, *23*, 1497.

[15] Wang, Z.-X.; Zhang, W.; Wu, C.; Lei, H.; Cieplak, P.; Duan, Y. *J. Comput. Chem.* 2006, *27*, 781.

[16] Meng, E. C.; Cieplak, P.; Caldwell, J. W.; Kollman, P. A. *J. Am. Chem. Soc.* 1994, *116*, 12061.

[17] Duan, L. L.; Mei, Y.; Zhang, D.; Zhang, Q. G.; Zhang, J. Z. H. *J. Am. Chem. Soc.* 2010, *132*, 11159.

[18] Duan, Y.; Wu, C.; Chowdhury, S.; Lee, M. C.; Xiong, G.; Zhang, W.; Yang, R.; Cieplak, P.; Luo, R.; Lee, T.; Caldwell, J.; Wang, J.; Kollman, P. *J. Comput. Chem.* 2003, *24*, 1999.

[19] Best, R. B.; Buchete, N.-V.; Hummer, G. *Biophys. J.* 2008, *95*, L07.

[20] Cieplak, P.; Dupradeau, F.-Y.; Duan, Y.; Wang, J. *J. Phys. Condens. Matter* 2009, *21*, 333102.

[21] Banks, J. L.; Kaminski, G. A.; Zhou, R.; Mainz, D. T.; Berne, B. J.; Friesner, R. A. *J. Chem. Phys.* 1999, *110*, 741.

[22] Rick, S. W.; Stuart, S. J.; Berne, B. J. *J. Chem. Phys.* 1994, *101*, 6141.

[23] Lamoureux, G.; MacKerell, A. D.; Roux, B. t. *J. Chem. Phys.* 2003, *119*, 5185.

[24] Vorobyov, I. V.; Anisimov, V. M.; MacKerell, A. D. *J. Phys. Chem. B* 2005, *109*, 18988.

[25] Ren, P.; Ponder, J. W. *J. Phys. Chem. B* 2003, *107*, 5933.

[26] Wang, J.; Cieplak, P.; Li, J.; Wang, J.; Cai, Q.; Hsieh, M.; Lei, H.; Luo, R.; Duan, Y. *J. Phys. Chem. B* 2011, *115*, 3100.

[27] Wang, J.; Cieplak, P.; Li, J.; Hou, T.; Luo, R.; Duan, Y. *J. Phys. Chem. B* 2011, *115*, 3091.

[28] Wang, J.; Cieplak, P.; Cai, Q.; Hsieh, M.-J.; Wang, J.; Duan, Y.; Luo, R. *J. Phys. Chem. B* 2012, *116*, 7999.

[29] Wang, L.-P.; Head-Gordon, T.; Ponder, J. W.; Ren, P.; Chodera, J. D.; Eastman, P. K.; Martinez, T. J.; Pande, V. S. *J. Phys. Chem. B* 2013, *117*, 9956.

[30] Gresh, N.; Cisneros, G. A.; Darden, T. A.; Piquemal, J.-P. *J. Chem. Theory Comput.* 2007, *3*, 1960.

[31] Xie, W.; Orozco, M.; Truhlar, D. G.; Gao, J. *J. Chem. Theory Comput.* 2009, *5*, 459.

[32] Gordon, M. S.; Smith, Q. A.; Xu, P.; Slipchenko, L. V. *Annu. Rev. Phys. Chem.* 2013, *64*, 553.

[33] Piquemal, J.-P.; Cisneros, G. A.; Reinhardt, P.; Gresh, N.; Darden, T. A. *J. Chem. Phys.* 2006, *124*, 104101.

[34] Ji, C.; Mei, Y.; Zhang, J. Z. H. *Biophys. J.* 2008, *95*, 1080.

[35] Duan, L. L.; Gao, Y.; Mei, Y.; Zhang, Q. G.; Tang, B.; Zhang, J. Z. H. *J. Phys. Chem. B* 2012, *116*, 3430.

[36] Ji, C. G.; Zhang, J. Z. H. *J. Am. Chem. Soc.* 2008, *130*, 17129.

[37] Ji, C. G.; Zhang, J. Z. H. *J. Phys. Chem. B* 2009, *113*, 13898.

[38] Ji, C. G.; Zhang, J. Z. H. *J. Phys. Chem. B* 2011, *115*, 12230.

[39] Li, Y.; Ji, C.; Xu, W.; Zhang, J. Z. H. *J. Phys. Chem. B* 2012, *116*, 13368.

[40] Zeng, J.; Jia, X.; Zhang, J. Z. H.; Mei, Y. *J. Comput. Chem.* 2013, *34*, 2677.

[41] Ji, C. G.; Zhang, J. Z. H. *J. Phys. Chem. B* 2009, *113*, 16059.

[42] Lu, Y.; Mei, Y.; Zhang, J. Z. H.; Zhang, D. *J. Chem. Phys.* 2010, *132*, 131101.

[43] Wei, C.; Tung, D.; Yip, Y. M.; Mei, Y.; Zhang, D. *J. Chem. Phys.* 2011, *134*, 171101.

[44] Xu, Z.; Lazim, R.; Sun, T.; Mei, Y.; Zhang, D. *J. Chem. Phys.* 2012, *136*, 135102.

[45] Gao, Y.; Lu, X.; Duan, L.; Zhang, D.; Mei, Y.; Zhang, J. Z. H. *Appl. Phys. Lett.* 2013, *102*, 193706.

[46] Li, Y. L.; Mei, Y.; Zhang, D. W.; Xie, D. Q.; Zhang, J. Z. H. *J. Phys. Chem. B* 2011, *115*, 10154.

[47] Tong, Y.; Mei, Y.; Li, Y. L.; Ji, C. G.; Zhang, J. Z. H. *J. Am. Chem. Soc.* 2010, *132*, 5137.

[48] Xiang, Y.; Duan, L.; Zhang, J. Z. H. *Phys. Chem. Chem. Phys.* 2010, *12*, 15681.

[49] Ji, C. G.; Zhang, J. Z. H. *J. Comput. Chem.* 2012, *33*, 1416.

[50] Jia, X.; Zeng, J.; Zhang, J. Z. H.; Mei, Y. *J. Comput. Chem.* 2014, *35*, 737.

[51] Mei, Y.; Li, Y. L.; Zeng, J.; Zhang, J. Z. H. *J. Comput. Chem.* 2012, *33*, 1374.

[52] Yao, X. X.; Ji, C. G.; Xie, D. Q.; Zhang, J. Z. H. *J. Comput. Chem.* 2013, *34*, 1136.

[53] Gao, Y.; Lu, X.; Duan, L. L.; Zhang, J. Z. H.; Mei, Y. *J. Phys. Chem. B* 2012, *116*, 549.

[54] Bayly, C. I.; Cieplak, P.; Cornell, W.; Kollman, P. A. *J. Phys. Chem.* 1993, *97*, 10269.

[55] Cornell, W. D.; Cieplak, P.; Bayly, C. I.; Kollmann, P. A. *J. Am. Chem. Soc.* 1993, *115*, 9620.

[56] Cieplak, P.; Cornell, W. D.; Bayly, C.; Kollman, P. A. *J. Comput. Chem.* 1995, *16*, 1357.

[57] van der Vaart, A.; Bursulaya, B. D.; Brooks, C. L.; Merz, K. M. *J. Phys. Chem. B* 2000, *104*, 9554.

[58] Soto, P.; Mark, A. E. *J. Phys. Chem. B* 2002, *106*, 12830.

[59] Pantoja-Uceda, D.; Pastor, M. T.; Salgado, J.; Pineda-Lucena, A.; Pérez-Payá, E. *J. Pept. Sci.* 2008, *14*, 845.

[60] Nymeyer, H. *J. Chem. Theory Comput.* 2008, *4*, 626.

[61] Onufriev, A.; Bashford, D.; Case, D. A. *Proteins: Struct. Funct. Bioinform.* 2004, *55*, 383.

[62] Kumar, S.; Rosenberg, J. M.; Bouzida, D.; Swendsen, R. H.; Kollman, P. A. *J. Comput. Chem.* 1992, *13*, 1011.

[63] Chodera, J. D.; Swope, W. C.; Pitera, J. W.; Seok, C.; Dill, K. A. *J. Chem. Theory Comput.* 2006, *3*, 26.

[64] Wu, Y.-D.; Zhao, Y.-L. *J. Am. Chem. Soc.* 2001, *123*, 5313.

[65] Viswanathan, R.; Asensio, A.; Dannenberg, J. J. J. Phys. Chem. A 2004, 108, 9205.

[66] Kobko, N.; Dannenberg, J. J. J. Phys. Chem. A 2003, 107, 10389.

[67] Ireta, J.; Neugebauer, J.; Scheffler, M.; Rojo, A.; Galván, M. J. Phys. Chem. B 2003, 107, 1432.